INTERNATIONAL UNION OF CRYSTALLOGRAPHY
BOOK SERIES

IUCr Monographs on Crystallography

1 *Accurate molecular structures*
A. Domenicano, I. Hargittai, editors
2 *P.P. Ewald and his dynamical theory of X-ray diffraction*
D.W.J. Cruickshank, H.J. Juretschke, N. Kato, editors
3 *Electron diffraction techniques, Vol. 1*
J.M. Cowley, editor
4 *Electron diffraction techniques, Vol. 2*
J.M. Cowley, editor
5 *The Rietveld method*
R.A. Young, editor
6 *Introduction to crystallographic statistics*
U. Shmueli, G.H. Weiss
7 *Crystallographic instrumentation*
L.A. Aslanov, G.V. Fetisov, J.A.K. Howard
8 *Direct phasing in crystallography*
C. Giacovazzo
9 *The weak hydrogen bond*
G.R. Desiraju, T. Steiner
10 *Defect and microstructure analysis by diffraction*
R.L. Snyder, J. Fiala and H.J. Bunge
11 *Dynamical theory of X-ray diffraction*
A. Authier
12 *The chemical bond in inorganic chemistry*
I.D. Brown
13 *Structure determination from powder diffraction data*
W.I.F. David, K. Shankland, L.B. McCusker, Ch. Baerlocher, editors
14 *Polymorphism in molecular crystals*
J. Bernstein

15 *Crystallography of modular materials*
G. Ferraris, E. Makovicky, S. Merlino

16 *Diffuse X-ray scattering and models of disorder*
T.R. Welberry

17 *Crystallography of the polymethylene chain: an inquiry into the structure of waxes*
D.L. Dorset

18 *Crystalline molecular complexes and compounds: structure and principles*
F.H. Herbstein

19 *Molecular aggregation: structure analysis and molecular simulation of crystals and liquids*
A. Gavezzotti

20 *Aperiodic crystals: from modulated phases to quasicrystals*
T. Janssen, G. Chapuis, M. de Boissieu

21 *Incommensurate crystallography*
S. van Smaalen

22 *Structural crystallography of inorganic oxysalts*
S.V. Krivovichev

23 *The nature of the hydrogen bond: outline of a comprehensive hydrogen bond theory*
G. Gilli, P. Gilli

24 *Macromolecular crystallization and crystal perfection*
N.E. Chayen, J.R. Helliwell, E.H. Snell

IUCr Texts on Crystallography

1 *The solid state*
A. Guinier, R. Julien

4 *X-ray charge densities and chemical bonding*
P. Coppens

7 *Fundamentals of crystallography, second edition*
C. Giacovazzo, editor

8 *Crystal structure refinement: a crystallographer's guide to SHELXL*
P. Müller, editor

9 *Theories and techniques of crystal structure determination*
U. Shmueli

10 *Advanced structural inorganic chemistry*
Wai-Kee Li, Gong-Du Zhou, Thomas Mak

11 *Diffuse scattering and defect structure simulations: a cook book using the program DISCUS*
R.B. Neder, T. Proffen

12 *The basics of crystallography and diffraction, third edition*
C. Hammond

13 *Crystal structure analysis: principles and practice, second edition*
W. Clegg, editor

14 *Crystal structure analysis: a primer, third edition*
J.P. Glusker, K.N. Trueblood

Macromolecular Crystallization and Crystal Perfection

NAOMI E. CHAYEN

Biomolecular Medicine, Department of Surgery and Cancer, Imperial College London, UK

JOHN R. HELLIWELL

School of Chemistry, University of Manchester, UK

EDWARD H. SNELL

Hauptman-Woodward Medical Research Institute, Buffalo, USA

OXFORD
UNIVERSITY PRESS

OXFORD
UNIVERSITY PRESS

Great Clarendon Street, Oxford OX2 6DP

Oxford University Press is a department of the University of Oxford.
It furthers the University's objective of excellence in research, scholarship,
and education by publishing worldwide in

Oxford New York

Auckland Cape Town Dar es Salaam Hong Kong Karachi
Kuala Lumpur Madrid Melbourne Mexico City Nairobi
New Delhi Shanghai Taipei Toronto

With offices in

Argentina Austria Brazil Chile Czech Republic France Greece
Guatemala Hungary Italy Japan Poland Portugal Singapore
South Korea Switzerland Thailand Turkey Ukraine Vietnam

Oxford is a registered trade mark of Oxford University Press
in the UK and in certain other countries

Published in the United States
by Oxford University Press Inc., New York

First published 2010

British Library Cataloguing in Publication Data
Data available

Library of Congress Cataloging in Publication Data
Data available

Typeset by SPI Publisher Services, Pondicherry, India

ISBN 978-0-19-921325-2 (Hbk.)

Contents

PART III DIFFRACTION

Preface

Structural crystallography provides key information to understand biological mechanisms. To this end the technique requires high-quality crystals which are necessary for determining the structure and dynamics of biological macromolecules.

These structural objectives have required a concerted action across the field of macromolecular crystallography over the last several decades and embraced various science disciplines. Notably these have included molecular biology for sample preparation and molecular engineering; a science and understanding for macromolecular crystal-growth conditions; applied physics for improved X-ray and neutron sources as well as detectors and for a characterization of the crystal perfection and harnessing white-beam diffraction and anomalous scattering in data collection; software for numerous stages of calculation; structural chemistry and biology for the basis of enzymatic function and the mechanisms of disease; and finally systems biology to gather these results into a unified genomic framework of interacting proteins and nucleic acids.

The methods today have a sense of maturity; after all, the Protein Data Bank is on target for 100 000 macromolecular crystal structures within a few years from now. However, the motivation for this book stems from the fact that the field is still afresh with change which arises for various reasons:

- the field continues to expand and new investigators and students are turning to macromolecular crystallography for tackling critical biological and chemical problems, often without training in physics in general or methods in particular, and who we aim to serve in providing in one volume the key topics of macromolecular crystallization and crystal use in diffraction structure determination;
- the sources of X-ray and neutrons, and associated apparatus, are still changing radically, bringing new capabilities;
- the understanding of crystal perfection as well as types of disorders and how to deal with them is improving; the types of twinning and lattice disorders of crystals is an excellent example and where protocols for remediation in the most awkward cases are becoming known; there are more adventurous diffraction geometries for protein structure determination being developed including protein powder X-ray diffraction and single-molecule structure determination envisaged with the X-ray lasers that are coming on line.
- synchrotron X-ray microbeams will soon be routinely available to be applied to micrometre and probably sub-micrometre crystals.
- the pixellated area detectors have recently arrived after many years of development and will allow shutterless X-ray data collection as the norm;

- consideration is now being made to archive raw diffraction images. This would provide the most complete record for a given crystal's diffraction characteristics rather than the merged, unique, Bragg diffraction 'spot intensities or amplitudes' alone. An added benefit of this approach is that the information on the continuum diffuse scattering would be retained until such time as its interpretation might become fully possible thereby giving further insight into molecular flexibility and motion;
- this book interfaces with the complementary techniques of structure determination which include solution small-angle X-ray scattering (SAXS), and cryo–electron crystallography, microscopy and tomography applied to large multi-macromolecular complexes.

We hope we have provided the reader with an overview of the complex topics involved in structural crystallography together with some of the more esoteric areas that can become very critical in key cases.

Naomi E. Chayen
London, UK

John R. Helliwell
Manchester, UK

Edward H. Snell
Buffalo, USA

Acknowledgements

Professor Chayen gratefully acknowledges the continuous support from the UK Engineering and Physical Sciences Research Council (EPSRC) and the European Commission OptiCryst Project. Previous support from BBSRC and the Leverhulme Trust are also acknowledged. Professor Helliwell is pleased to acknowledge all of his collaborators, PhD students and post-doctoral workers who have contributed directly or indirectly to these research efforts. He thanks the Universities of Manchester, York, Keele, Oxford and the STFC Daresbury Laboratory for general support, excellent facilities and stimulating research environments over many years stretching back over the decades and including his DPhil period. He is pleased to acknowledge the UK research funding agencies, successively SERC then BBSRC and EPSRC and the MRC; The Wellcome Trust; the EC; The Leverhulme Trust; the Nuffield Foundation; the British Council; the Hasselblad Foundation and The Royal Society as well as the Swedish Research Council and the European Space Agency (ESA). Dr Snell would like to acknowledge the help and support of the staff at the Hauptman-Woodward Medical Research Institute in particular Dr George DeTitta, Joseph Luft, Thomas Grant, and the High-throughput Crystallization Screening Laboratory. He would also like to acknowledge NIH and the Oishei Foundation for their support as well as NASA. Professors Chayen and Helliwell acknowledge research grant support from the EC under the European Biocrystallogenesis Initiative (EBCI). We would like to acknowledge the following synchrotron radiation facilities: SRS, ESRF, CHESS, APS, NSLS, SSRL and DLS as well as the Institut Laue Langevin neutron research centre where much of our macromolecular crystallography research has been conducted.

We are grateful and have benefitted from extensive comments by J. Luft (crystallization and diagnostics), as well as from M. Nespolo and A. Lebedev (twinning) and J. Spence (X-ray laser applications). We are also grateful to all respondents (corresponding authors and Journals) for their prompt attention to our copyright permissions requests and their help with occasional detailed rewording of associated text, Table and/or figure captions. Any errors that remain are of course our own.

Cover picture. A montage showing clockwise from top left; a typical monochromatic X-ray diffraction pattern from a macromolecular crystal, crystals of the C1 Domain of Cardiac Myosin Binding Protein-C (Govada and Chayen, 2009) (grown by Dr Lata Govada, Imperial College London), Laue neutron diffraction pattern (also from a macromolecular crystal), the hutch at beamline 11-1 at SSRL (courtesy of Dr Aina Cohen, SSRL, Stanford), the ESRF and ILL facilities in Grenoble France (courtesy Denis Morel, ESRF, Grenoble), and an example of the ultimate quality data that can be achieved, an ultrahigh-resolution electron-density image.

PART I

INTRODUCTION AND OVERVIEW

1

Introduction

Biological macromolecules are the machinery of life; visualizing their three-dimensional structure helps us to fully understand their function. Visual observation with a light microscope is not possible as their sizes are well below the wavelength of visible light. While X-rays and neutrons allow visualization, they cannot be focused, so diffraction techniques have to be used. An understanding of three-dimensional macromolecular structure gives us a deeper understanding of basic biological concepts and processes, reveals the causes of diseases, assists rational pharmaceutical design and can lead to the design of macromolecules with novel properties. Visualizing these macromolecules is a complex ballet involving diverse but interrelated fields of endeavour. In this book, we aim to describe in some detail these complementary techniques, which include crystallization, diffraction and analysis of the data to obtain atomic structure from crystals of macromolecules. We cover areas where problems can occur and potential solutions to those problems. Finally, we touch on some of the developments in the not so distant future.

When we use the term crystal in the context of a biological macromolecule, we are describing an ordered array of macromolecules in an environment that keeps them stable. Biological macromolecules are predominantly made up of low atomic weight atoms, including hydrogen, carbon, nitrogen, oxygen, sulphur and phosphorous. Unlike inorganic crystals, a significant proportion (30–70%) of a macromolecular crystal is water (Matthews, 1968). This makes biological crystallography challenging; the process of crystallization is very complex and the crystals themselves diffract very weakly in comparison to inorganic crystals due to the low atomic weight content and disordered water making up a large proportion of the crystal volume. The biochemical and biophysical process of crystallization is still a largely empirical process. High-throughput technologies have been employed to speed up the crystallization process, allowing for more experiments to be set up using less sample, but their greater power may be realized if we achieve the array of experiments needed to better understand this complex process to develop crystallization.

Once we get the crystals, technologies and methods have advanced significantly and transformed our capabilities for structure analysis. In the X-ray field, highlights of the last three decades have included synchrotron radiation, detectors and cryo-crystallography. In a synergistic development neutron Laue methods, along with very large area neutron-sensitive image plates, and new spallation source developments, are enhancing the potential for complete, i.e. with-H atoms (as deuteriums), structures. Complementary to these are developments in molecular biology techniques that go beyond the simple cloning and expression (production) of the target sample. Rational mutagenesis is providing functional information and being used to improve crystal quality, while fully perdeuterated protein production is enabling neutron studies.

1.1 Crystal Growth

The history of crystallization has been described by McPherson (1991; 1999a). The first published observation of crystallization, haemoglobin, was noted by Hünefeld (1840) when the blood of an earthworm was pressed between two microscope slides. This was followed by a slow growth in a number of other crystallized macromolecules until the 1930s when the pace quickened. Crystallization was initially used as a method of isolation with Northrop, Sumner and Stanley sharing the 1946 Nobel Prize for chemistry for the isolation and crystallization of proteins and viruses. The first biomolecular crystal-structure determined was that of vitamin B-12 in 1957 by Dorothy Crowfoot Hodgkin (1957) who subsequently won the Nobel prize for chemistry as a result of the work. Initially there were no set rules or recipes as to where to start to crystallize a macromolecule. Macromolecules were solubilized and then treated with arrays of precipitants in order to find favourable conditions for crystallization.

Crystallization is still largely empirical with many experiments sampling a large range of possible crystallization conditions. This idea of a designed sampling of many conditions, or screening, was introduced in 1979 (Carter and Carter). The numerous experiments required repetitive pipetting, a laborious, time-consuming and tedious task but the principal difficulty was attributing a quantitative score to the results to enable a meaningful mathematical analysis. In the late 1980s and early 1990s, the development of automatic means of dispensing crystallization trials (Chayen *et al.*, 1992; Chayen *et al.*, 1994; Chayen *et al.*, 1990; Cox and Wever, 1987; Oldfield *et al.*, 1991; Rubin *et al.*, 1991; Sadaoui *et al.*, 1994; Soriano and Fontecillacamps, 1993; Ward *et al.*, 1988) showed the promise of designed screening. However, it was not until the commercialization of a crystallization screen developed by Jancarik and Kim (1991) that systematic screening became a standard laboratory technique.

In the 1980s an effort was initiated to turn crystallization from an art into a science; the first of a continuing series of international conferences on the subject occurred (McPherson and Giege, 2007) and the term 'crystallogenesis' was coined (Giege *et al.*, 1986). The purpose of crystallogenesis was to understand the fundamental principles of the crystallization process, to quantitatively measure the biophysical and chemical parameters that are involved in crystal growth, and to use that knowledge to design experiments for obtaining better diffracting crystals. Systematic studies were performed mostly on lysozyme and other model proteins that crystallized with ease.

The aim was to extrapolate that information to target proteins that were proving difficult to crystallize. Progress was slow due to lack of suitable equipment for monitoring the crystallization process and to the complexity of the problem. Even the crystallization process for model proteins was not straightforward given the number of variables that were involved. Considerable development effort has been ongoing in the last 20 years on crystal-growth methods as well as more sophisticated monitoring and characterization, such as use of light scattering, interferometry and other techniques.

Much of the success of the last decade has come by way of automating and miniaturizing crystallization trials (Kuhn *et al.*, 2002; Luft *et al.*, 2003; Walter *et al.*, 2003) and by way of the development of diagnostic apparatus to study the crystallogenesis aspect of crystallization (Dierks *et al.*, 2008; Yeh and Beale, 2007). The ability to dispense trials consisting of nanolitre volumes in a high-throughput mode has cut the time of setting up experiments from weeks to minutes, and reduced sample requirements by an order of magnitude, a scenario that was unimaginable even in the recent past. While high-throughput approaches and miniaturization do not elicit a better understanding of crystallization, the analysis of these systematic and highly reproducible trials will improve our comprehension of the crystallization process, and application of these methods to specifically understand this process will enable us to answer and then ask many more questions.

Having a well-diffracting single crystal is a first step, but is not necessarily sufficient to solve the macromolecular structure. Detectors are able to measure the position and intensity (amplitude) of scattered reflections, but do not record phase information. In order to use a Fourier transform to go from a diffraction pattern, to an interpretable electron-density map that can be used to model the structure, we need phase information. We can use several approaches to provide an initial set of phases. If the protein has fewer than 1,000 non-hydrogen atoms, and the resolution of the diffraction data is near atomic resolution, we can use *ab initio* phasing (direct methods) to solve the structure (Hauptman, 1997; Uson and Sheldrick, 1999), or if there is significant sequence homology to other known structures, then molecular replacement may be used to solve the structure without modifying the sample.

If not, sample modification will likely be required. This can be accomplished by soaking, or co-crystallizing the sample with heavy atoms to provide phase information for a sub-structure (Islam *et al.*, 1998). Another approach, exploits differences in diffraction intensities caused by anomalous scattering, absorbance of X-rays by elements at wavelengths at, or near a particular element's absorption edge. The most common application of this method uses molecular biology to replace naturally occurring sulphur atoms in the protein's methionine residues with selenium atoms, creating a selenomethionine variant (Hendrickson *et al.*, 1990). The derivative is crystallized, and diffraction data collected at several wavelengths near the selenium absorption edge; this is an example of MAD (multiple anomalous dispersion) phasing. SAD (single anomalous dispersion) phasing is similar, but uses only one wavelength, making this a better-suited method for radiation-sensitive crystals (Gonzalez, 2007).

A special aspect of this history is the growth of macromolecular crystals for neutron crystallography. Neutrons are weakly scattered and neutron fluxes are low compared to X-ray sources. Before the advent of area detectors at neutron sources and the use of polychromatic beams, this challenge was particularly acute, whereby crystal volumes on the order of $\sim$30 mm^3 were required and that measurable diffraction often could not be obtained within a reasonable timeframe. To achieve these large crystals, a crystal of 'typical size' would often be transferred into fresh growth solution (Lennart Sjolin, personal communication). Specialized image plates at reactor neutron sources have made a big impact on the efficiency of data collection compared to the point-by-point or linear detector diffractometers. The use of Laue techniques, e.g. a wavelength bandpass of 2.5 to 3.5 Å has increased the efficiency of measuring many reflections at once and also helped by allowing such long wavelengths to enhance the scattering efficiency for each spot via a λ^2 effect.

Production of fully deuterated protein for neutron protein crystallography, initiated by Gamble *et al*., (1994) has led to a further reduction in crystal sample volumes due to two effects. First, for coherent neutron scattering, that produces Bragg reflections, deuterium scatters neutrons as strongly as carbons; there is a doubling of the strongly scattering atoms since there are so many hydrogens in a protein. Secondly, the incoherent scattering cross-section of hydrogen is anomalously large, causing high background scattering from hydrogen. This is reduced significantly when hydrogen is replaced by deuterium. The combination of these two factors provides for nearly an order of magnitude improvement in the neutron protein crystallography signal-to-noise ratio (Myles, 2006).

Finally, source developments help neutron diffraction. The new spallation neutron sources offer peak flux improvements and time-of-flight measurements that allow the background being accumulated over a diffracted spot to be kept to the minimum. All these developments have radically reduced the required crystal volume but it is still much larger than that required for X-rays.

While neutrons have required larger crystals, developments in intense and tunable X-ray beams have steadily allowed smaller and smaller crystal volumes to produce useable data with now down to micrometre sized beams and sample volumes. Plans for future upgrades and beamlines (e.g. the ESRF Upgrade) offer sub-micrometre and indeed 'nano-' sized X-ray beams for possible further reductions in sample volume.

1.2 Diffraction Techniques

1.2.1 X-rays

X-rays were discovered in 1895 by Röntgen and in 1912 Friedrich, Knipping and von Laue discovered that they were diffracted by crystals of small molecules. There is a large literature available on the development of X-ray single-crystal techniques and that literature, critical to understanding how crystal quality is important, is summarized here. For the oscillation method in single-crystal X-ray diffraction the integrated intensity of a reflection (h,k,l) for atoms at rest, $E(h,k,l)$, can be calculated from:

$$E(h,k,l) = \frac{\lambda^3}{\omega . V^2}\left(\frac{e^2}{mc^2}\right)^2 V_{cr} I_o LPA|F(h,k,l)|^2 \qquad (1.1)$$

where λ is the wavelength, ω the angular rotation velocity, V the unit-cell volume, V_{cr} the illuminated volume of the crystal, I_o the incident intensity, L the Lorentz factor, P the polarization, A the absorption and F the structure factor (Woolfson, 1997). The structure factor is given by

$$F(h,k,l) = V \int_{\text{cell}} \rho(x,y,z) e^{[2\pi i(hx+ky+lz)]} dx\, dy\, dz \qquad (1.2)$$

where x, y and z are the fractional coordinates in the unit cell and $\rho(x,y,z)$ is the electron density at those coordinates. From this we can see how diffraction intensity decreases as the square of the unit-cell volume; large macromolecules or complex systems will diffract poorly compared to smaller counterparts. The dependence on wavelength can be misleading as absorption and practical considerations may limit signal improvements that can be gained by changing wavelength.

All contemporary laboratory sources operate on a principle of electrons striking an anode with radiation given off from the anode at a characteristic wavelength associated with the anode material, for example Cu=1.54 Å, Co=1.80 Å, *etc*.

Synchrotrons have been one of the most dramatic developments in X-ray crystallography. Electrons (or positrons) are accelerated to high velocity and stored in a circular orbit. As they lose energy, the stored particles give off electromagnetic energy, some of it in the X-ray spectrum. The first synchrotrons started life as experimental machines for particle physics and were used by crystallographers in a parasitic mode; the loss of energy was considered detrimental. These synchrotrons were superseded by a second generation of machines designed and dedicated solely to provide synchrotron radiation for experimental use. The current third-generation synchrotron source are machines designed to incorporate insertion devices.

1.2.2 Neutrons

Neutrons were discovered later than X-rays, by Chadwick in 1932, and the first diffraction experiments took place in 1945 by Wollan. Neutrons, produced in a nuclear reactor, are slowed down such that their wavelength is similar to that of X-rays used for crystallography. X-rays interact with the electron cloud, while neutrons interact with the nucleus of the atom. In the case of X-rays, the X-rays scattered are, to a first approximation, proportional to the atomic number. For neutrons the contribution to the scattered intensity is different for each isotope, illustrated in Figure 1.1. In macromolecular crystallography we can make use of this to differentiate hydrogen from deuterium or to examine the protonation state of an amino acid residue (see e.g. Shu *et al.*, (2000)). Neutrons are scattered by the nucleus causing relatively little

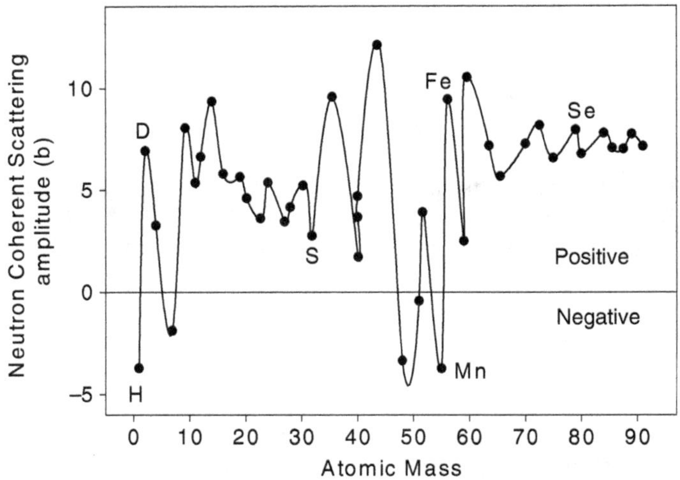

Fig. 1.1. Plot of neutron coherent scattering amplitude as a function of atomic number.

decrease in intensity with scattering angle, unlike X-rays that are scattered by the much larger electron charge cloud. Obviously ADPs (atomic displacement factors) (deleteriously) affect the fall off with scattering angle in both the neutron and X-ray cases. Neutrons do not cause radiation damage or suffer absorption, allowing ready use of larger crystals.

The original neutron source, the nuclear reactor, has also been joined by spallation sources. When a high-energy proton bombards a heavy atom nucleus some neutrons are 'spalled' or knocked out. For every proton hitting a nucleus some 20 to 30 neutrons are expelled.

Neutron sources have relatively low flux compared to synchrotron X-ray beamlines, in fact typical sources have a flux comparable to that of an X-ray tube. In addition, the scattering of neutrons is weak. To enable optimum experimental results, the neutron flux and diffraction signal have to be maximized and the noise minimized. Table 1.1 (Snell *et al.*, 2006) summarizes several complementary pathways to achieve this goal with a qualitative estimate of the relative cost, time and chance of success.

Certain approaches, i.e. increasing the neutron flux or improving the detection of the signal, are beyond the resources of an individual user and are being addressed at the facility level, e.g. in the construction of dedicated instruments on new neutron sources such as the Spallation Neutron Source (SNS), USA, and the Japan Proton Accelerator Complex (J-PARC). Other techniques, for example improving the signal-to-noise ratio by perdeuteration (replacing hydrogen with deuterium atoms) may be routinely attempted in the laboratory (Hazemann *et al.*, 2005; Shu *et al.*, 2000) and can dramatically decrease the volume of the crystal needed for successful data collection (Hazemann *et al.*, 2005; Shu *et al.*, 2000).

Table 1.1. Summary of different approaches to improve reflection intensity from neutron diffraction divided into increasing neutron flux and brilliance, improving the instrument, reducing the noise or improving the sample (beyond optimization in the crystallization step). All the approaches are complementary and cumulative. They are assessed by the relative cost, time and success of implementation (Snell *et al.*, 2006).

Solution	Relative cost	Time	Success
More neutrons			
Increase source intensity	Very expensive	Long term	Certain
Reduce distance to source	Expensive	Medium term	Certain
Increase exposure time	Inexpensive but reduces throughput	Short term	Certain
Improved detection/optics			
New/improved detector technology	Expensive	Medium term	Good
Focusing optics	Moderately expensive	Medium term	Good
Improve signal-to-noise			
Deuteration	Relatively inexpensive	Short term	Very good
Diffracting volume			
Grow larger crystals	Inexpensive	Short-term	Good

1.3 Crystal Volume and Quality

A high-quality crystal can be defined as one that provides structural information that is of sufficient detail (resolution) to answer the structural question being posed. A crystal of sufficient volume and internal order is required to measure complete, observable diffraction data at high resolution.

Examining this definition, the first requirement, crystal sample volume, is an easily quantified metric. For X-ray single-crystal studies structural information can currently be extracted from crystals as small as 20 μm in diameter (Hedman *et al.*, 1985; Pechkova and Nicolini, 2004a; Pechkova and Nicolini, 2004b). In the extreme case structural information can be extracted from microcrystallites or powder. For neutron diffraction the volume requirement is now falling below 1 mm^3 with perdeuteration techniques of the protein itself (Hazemann *et al.*, 2005) and now new neutron spallation sources coming on-line. Many crystal-growth methods and optimization techniques exist to address improving crystal volume.

While crystal volume can be determined visually, the requirement for order within the crystal can only be determined by diffraction methods. What determines order within a crystal? Ideally, well-ordered crystal macromolecules would be conformationally identical, down to the position of every sidechain on each amino acid residue. If a single sidechain on a sequentially identical amino acid residue holds two stationary positions we have what is termed a static disorder. An example of this situation is when two structural states are seen in the structural model derived from the electron-density map, rather than one ‘unique’ state. The split-occupancy values in favourable cases are refined against the Bragg reflection intensities. If one can see such 'split-occupancy' details in a crystal structure, e.g. of an amino acid sidechain in two states, then it can be viewed as a remarkable state of order each with their fractional occupancy (constrained to add to 100%). Furthermore, if one cannot see evidence of the structure at all in regions where it is clearly present then the term disordered makes the most descriptive sense. There is in such a way of describing it a logical progression here of fully ordered, partially ordered and finally disordered. However, the tradition has grown up of referring to cases of less than 100% order as 'static disorder', i.e. a crystal is termed imperfect even if, say, 40% of it is ordered one way and 60% in another. Another type of disorder is occupational, i.e. two or more elements occupying the same site. Further, aperiodic structures are those that are fully ordered in a higher dimension (e.g., a modulated structure or quasi-crystal).

The lattice scale internal crystal order can be divided into two distinct properties, short-range order and long-range order (Snell *et al.*, 2003). Disruptions to the internal crystal order lead to a 'bad' crystal with a number of associated symptoms;

- The crystal may only diffract to low resolution.
- The diffraction spots themselves have texture features and/or are basically split.
- There may be few or no spots even if the crystal looks optically well defined.
- There is a significant anisotropic local order variation within the crystal revealed by detailed examination of the diffraction data.
- There is a superlattice of closely spaced diffraction spots.
- Crystal twinning can be bad but is now increasingly amenable to detwinning diffraction data analysis computational procedures. Alternatively more crystal growth 'remediation methods' are now known.

To understand these symptoms we will look in detail at both short- and long-range order then the tools and techniques that can be brought to bear for a systematic exploration of order in terms of crystal quality.

1.3.1 Short-range order (intermolecular)

Good short-range order in a crystal is a primary needs-driver for the crystallographer where high-resolution structural information is the required feature. An atom will contribute coherently to the intensity of a reflection only if its disorder relative to symmetry-related atoms is small. Figure 1.2 dissects the various short-range intermolecular ***disorders*** that can occur on the molecular scale within a crystal. Atoms can be displaced by thermal vibrations; they can have multiple or partial occupancies; their position may be uncertain, especially in the case of waters and there may be variations in the main chain or sidechains and in the intermolecular packing. There may also be vacancies or interstitial substitutions, which clearly also interrupt crystal periodicity.

Atomic displacement or temperature effects

Temperature effects reduce the integrated reflection intensity as a function of resolution. This temperature-dependent atomic displacement parameter, *ADP*,

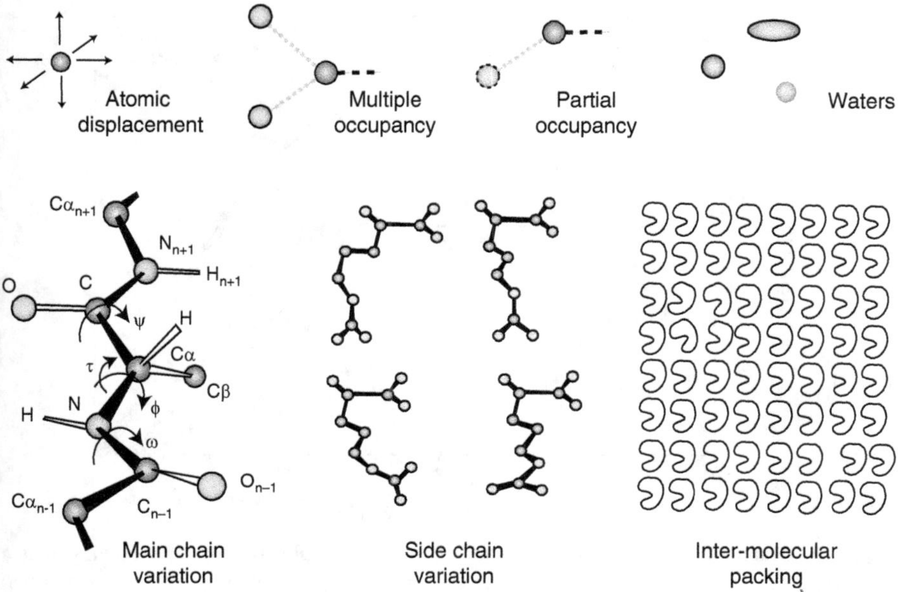

Fig. 1.2. Contributors to the reduction of short-range order within a macromolecular crystal.

often referred to as the B_{factor}, is related to the mean square atomic displacement by:

$$B_{\text{factor}} = ADP_i = 8\pi^2 < u_i^2 > \tag{1.3}$$

where $< u_i^2 >$ is the mean square atomic displacement of the atomic vibration for atom i. The overall temperature factor for the crystal, averaged over the whole structure, can be calculated from a Wilson plot (1942) where E_{obs} the observed intensity of the reflection, is plotted against $(\sin\theta/\lambda)^2$. The $<B_{\text{factor}}>$ is extracted from the slope of this plot as:

$$E_{\text{obs}}(h, k, l) = E_{(h,k,l)} e^{(-2<B_{\text{factor}}>\sin^2\theta/\lambda^2)} \tag{1.4}$$

The Wilson plot also provides a scale factor, where it crosses the vertical axis, allowing intensities to be put on an absolute scale.

The atomic-displacement parameter is a measure of how much dynamic disorder of an atom is within the crystal lattice on the molecular level. X-rays interact with the electron cloud, which depends on the atomic position. Each atom vibrates around an equilibrium position so the nuclear position and electron charge cloud both move. The ADP is an indication of both this motion around the equilibrium of the ith atom position **and** the deviation in position of that atom from molecule to molecule making up the crystal lattice. As can be seen from eqn (1.4), reducing the $<B_{\text{factor}}>$ has a

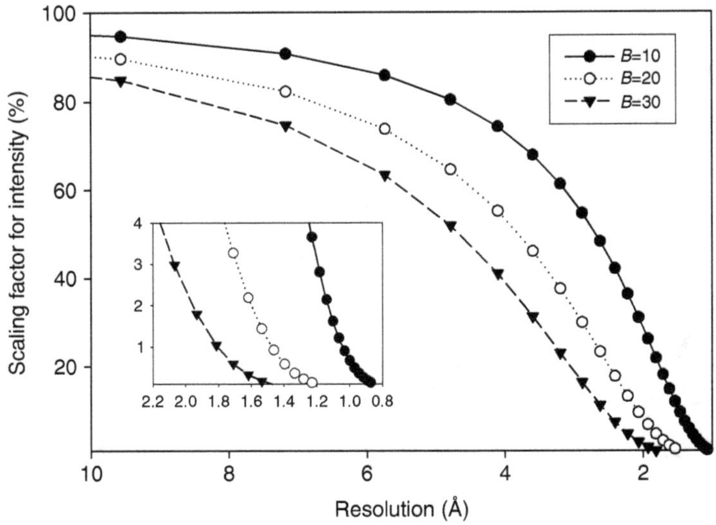

Fig. 1.3. Graph of scaling factor for intensity with resolution (eqn (1.5)) showing that a decrease in the $<B_{\text{factor}}>$ causes a significant increase in the resolution. This is a simplified example neglecting crystal path length absorption differences and Bragg reflection signal-to-noise smearing at higher resolutions.

dramatic effect on increasing the signal-to-noise of the data. A small ADP_i is indicative of a combination of small dynamic motion of the ith atom and good intermolecular order within the crystal. In Figure 1.3 the scaling factor $e^{(-2<B_{\text{factor}}>\sin^2\theta/\lambda^2)}$ is plotted against resolution for varying values of B. Increasing $<B_{\text{factor}}>$ results in a more rapid fall off of intensity as a function of resolution.

For a macromolecule, in practice, there is considerable shape to this plot and a curve rather than a straight line is seen at resolutions around 3 to 4 Å. This is due to the nature of macromolecular secondary structures, alpha helix and/or beta sheet, and their regularity, namely periodicity of the helix on the one hand and inter-strand separation on the other, causing the molecular transform to peak at these distances in reciprocal space. At higher resolution a straight line results and an accurate assessment of the $<B$ value$>$ becomes possible. The weakness of this parameter as the sole judge of optimal crystal-growth conditions is that it is only an average indicator summed over the innate flexibilities of the protein and the periodicity in a given crystal sample.

The relative Wilson plot has been used to evaluate the variation of crystal to crystal quality for crystals of the same protein either grown by different methods or where there is known to be variation within the same batch. This calculation has the advantage that the helix and sheet molecular effects then cancel out.

If we compare the protein molecule in a protein crystal situation in practice, described above, with a small-molecule crystal, then the distinctions regarding the $<B_{\text{factor}}>$ are as follows. A small-molecule crystal has, firstly, a generally much lower set of individual atomic ADP_i values and a generally better adherence to periodicity. Such molecules much less frequently have extensive mobile parts such as the loops of a protein and contain very little solvent of crystallization, unlike protein crystals. It used to be the case that whilst it was routine to cool a small-molecule crystal this was not done for a protein crystal. The latter has only become common practice since around the late 1980s, championed in a major way by the ribosome protein structure work of Ada Yonath with the cryogenic developments particularly of Hope *et al.* (1989), although conceived of in the 1950s (e.g. see King (1958)). Typically, a small-molecule crystal at room temperature has an overall B_{factor} of 6 and at cryogenic temperature (~100 K) of 3 whereas a protein crystal at room temperature has an overall $\langle B_{\text{factor}} \rangle$ in the range of 25 to 50 and at 100 K cryogenic temperature in the range of 8 to 25.

Variation in atomic positions or occupancy

The structure determined from X-ray diffraction data is then a time- and spatial-averaged structure of the complete number of individual macromolecules making up the crystal. This can be a large number. For example, if we consider a cubic crystal with dimensions of 100 μm made up of macromolecules with a unit cell of 100 Å on edge then there will be 10^{12} of those molecules in the crystal. An atomic displacement parameter takes into account the variation about a single atomic position. A number of residues may have alternate positions or multiple conformations. In some cases, e.g. in the interior of the macromolecule or at crystal lattice contacts, these conformations are clearly defined. In other cases, e.g. long flexible residues or complete loops on the outside of the protein, the atomic positions are not clearly defined. Multiple sidechain conformations can be determined in high-resolution analyses from about 1.5 Å onwards and the fraction of atoms in this category seems to be often more when the crystal structure has been determined at cryogenic rather than room temperature.

There are particularly long and therefore flexible amino acid sidechains such as lysine. The atoms in these sidechains have high atomic values of their ADP_i. Site-directed mutagenesis to replace such sidechains has become an approach to improve overall crystal quality and has enjoyed success (Derewenda, 2004a; Derewenda, 2004b) and is discussed in Chapter 6.

The bound-water structure on the surface of a protein can also reach around 10% of the X-ray crystal-structure determined atom positions in the cryo case and around 3 to 5% in the room-temperature case. The average ADP_is for this population of atoms is usually considerably more than the well-ordered protein atoms, e.g. twice as high but is a strong function of whether the bound water molecule is found in the first water layer or the second (a few waters can also be found in a third layer at cryo-temperature).

In the case of neutron protein crystallography where fully deuterated protein has been isolated and the bound water is D_2O the proportion of atoms that diffract as well as carbon atoms has risen considerably. This has a strong effect on increasing the diffraction signal. Moreover, the large incoherent background contribution of hydrogen in the interaction with a neutron beam is considerably reduced by replacement with deuterium in the protein and in the D_2O. Since deuterium scatters as well as carbon with neutrons then determination of the protonation states of ionizable amino acid sidechains is viable even at 2Å (or up to 2.3 Å or so) resolution. The use of X-rays, on the other hand, is vulnerable to the effects of specific atom ADP_i values, thus losing the relatively weak hydrogen scattering signal, even at atomic resolution, which is why neutrons can provide the critical complementary information even with a more modest diffraction resolution.

1.3.2 Long-range order (domain structure)

Long-range order is a whole-crystal length-scale effect. Good long-range order results in high signal-to-noise in the reflection profiles. The intensity of the reflection is directly proportional to illuminated volume, see eqn (1.1), but the peak width is also somewhat narrowed due to the larger volume, i.e. for an infinite perfect crystal the peak would be a Dirac delta function.

Figure 1.4 illustrates perturbations to long-range order and how long-range disorder contributes to broadening the resulting diffraction reflection profile. The figure makes use of the mosaic model of crystals proposed by Darwin (1922) and approximates the crystal to an array of perfectly ordered volumes (domains) slightly misaligned with respect to each other (the boundaries between these domains are ignored and no model for them is proposed). In addition to having small random misalignments, the domains can be of varying volume and to a much smaller degree the unit cells in the crystal can vary in dimension. Each of these phenomena has a distinct effect on the crystal (Boggon *et al.*, 2000; Nave, 1998). In the case shown in Figure 1.4(a) all the domains are well aligned so their contributions to the reflection overlap. Misalignment of the domains broadens the reflection profile reducing the signal-to-noise. If the volume of the domains becomes small, the reflections will become broadened from Fourier truncation effects (the transition from diffraction grating to a few slits is the analogous situation in optical diffraction and interference theory). The effect is known as domain-size broadening. A lattice parameter variation, Figure 1.4(c) causes a reflection to have a range of slightly different Bragg angles, also resulting in a smearing out of the reflection.

Long-range disorder in the crystal gives rise to localized effects in reciprocal space (Boggon *et al.*, 2000; Nave, 1998). Improved long-range order in a crystal reduces the mosaicity, which results in an increase in the signal-to-noise of the reflections. This is illustrated in Figure 1.5, where Gaussian profiles of the same integrated intensity but different full width at half-maximum values are shown. The increase in signal is directly proportional to the decrease in mosaicity.

Mosaicity is a global term covering contributions from all components shown in Figure 1.4. For practical purposes, it is defined as a function of the angular extent of the diffraction profile. In Chapter 11, the measurement of mosaicity and the

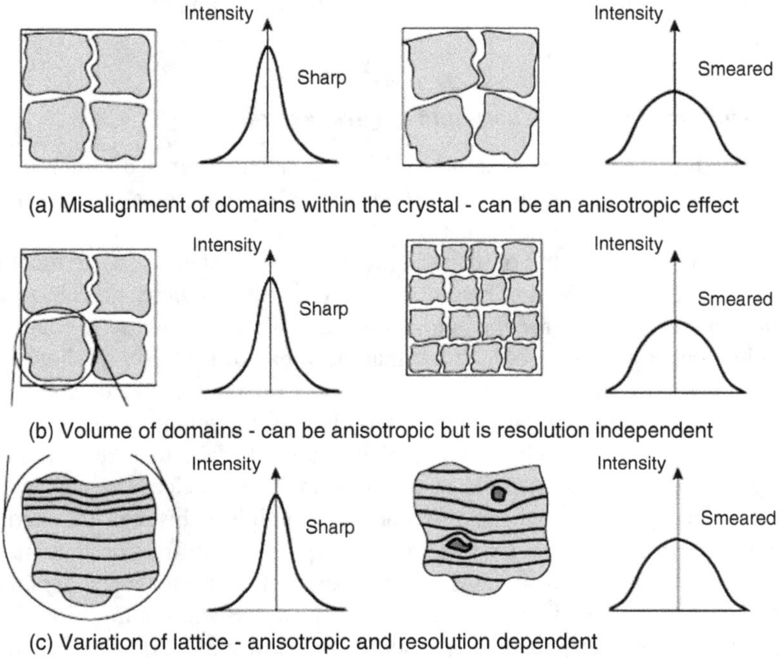

Fig. 1.4. Long-range disorder and the resulting effect on the diffraction profile.

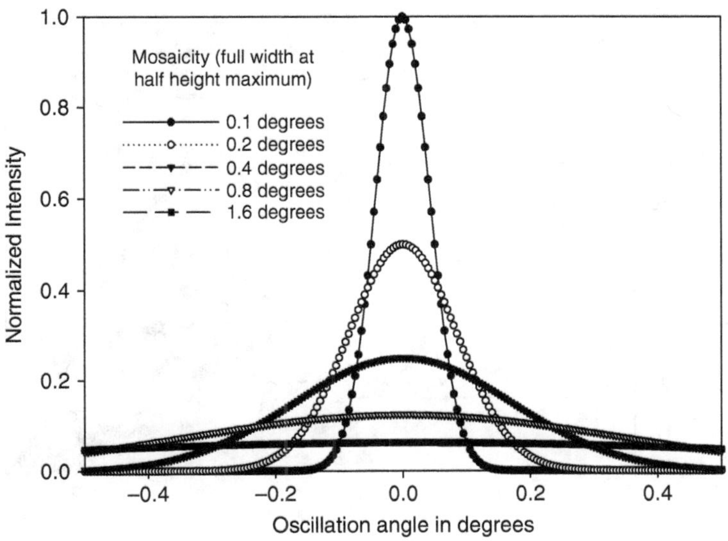

Fig. 1.5. Graph showing a Gaussian reflection profile for identical integrated intensities but differing full width at half-height maxima to illustrate how an increase in mosaicity causes a corresponding decrease in the reflection peak intensity signal.

deconvolution of it from experimental parameters and into the component parts is discussed.

1.3.3 The combination of short- and long-range order

Disorder within a crystal is a combination of short- and long-range disorder. In Figure 1.6, the scaling factor on intensity is plotted for a crystal having an overall $<B_{\text{factor}}>$ of 10 but mosaicity varying from 0.1 to 0.8 degrees.

Figure 1.6 shows a plot of the scaling factor for intensity as a function of resolution. The peak reflection height as a function of mosaicity has been used to calculate initial intensity for the plot. This provides a worst-case scenario since integration techniques are profile rather than peak based. However, it shows that in principle mosaicity has little effect on resolution. In practice, in crystal-structure analysis, the percentage of data with $I>2\sigma(I)$ is very important and is improved as Figure 1.6 shows nicely. How to optimize data collection to take advantage of reduced mosaicity is discussed later, but basically in practice account has to be taken of the background noise in a diffraction pattern. It is in this area that sharpening of mosaicity can have a positive impact on the quality of diffraction data measurement. Alternatively, of course the electronic gating of a detector to exclude noise is another strategy for achieving a signal-to-noise measurement optimization. In summary, reducing mosaicity in this case does not significantly extend the theoretical resolution but does dramatically increase the peak intensity of the signal. This can

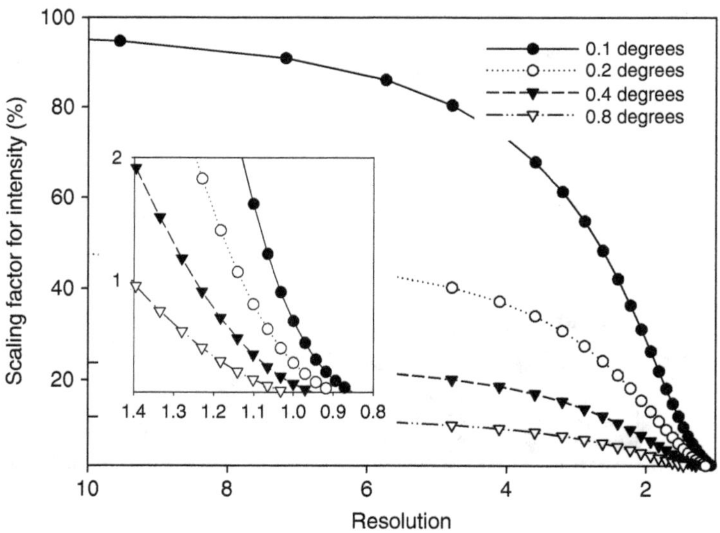

Fig. 1.6. Theoretical plot of scaling factor for intensity signal versus resolution for a crystal with mosaicity of 0.1 to 0.8 degrees. The plot is normalized against the 0.1 degree mosaicity crystal.

impact the resolution likely to be achieved in practice as measured by, e.g. the percentage of data with $I > 2\sigma(I)$.

1.4 Chapter Summary

To obtain structural data from crystallography crystals that diffract are needed. To obtain good data, those crystals have to be as perfect as possible. In the chapters ahead we explore crystallization going into theory and practical aspects. We discuss screening and optimization and how some of the high-throughput technologies developed for structural genomics efforts can be used to provide fundamental data on the science of crystallization. We discuss alternative strategies and salvage pathways if a crystal is not obtained or if the crystal that does result does not yield good enough structural data to resolve the question of interest. We shift from this to diffraction and what diffraction can reveal about the crystal quality and some of the problems associated with it. We address ways to improve the quality of diffraction data at this stage then look at rapidly developing areas where smaller crystals can be utilized, e.g. microfocus capabilities down to powder techniques. Finally, we end with a short note on complementary techniques and what we perceive to be on the near-term future horizon.

PART II

CRYSTALLIZATION

2

Crystallization theory

Crystallization is a phase-transition phenomenon. Crystals grow from an aqueous protein solution, when the solution is brought into supersaturation (Ataka, 1993; Ducruix and Giege, 1999; McPherson, 1999b). Supersaturation is achieved by surveying chemical makeup and concentrations of precipitant, protein and additives, and by varying other parameters known to affect solubility such as pH, and temperature (Chayen, 2004).

Crystallization proceeds in two phases, nucleation and growth. Nucleation is the conception stage of crystallization during which a molecular aggregate of critical size forms, allowing for subsequent crystal growth. A number of papers have studied the nucleation and its mechanism in detail (Chernov, 2003; Garcia-Ruiz, 2003b; McPherson *et al.*, 2001; Plomp *et al.*, 2003; ten Wolde and Frenkel, 1997; Yau and Vekilov, 2000). Nucleation and crystal growth both take place under conditions of thermodynamic supersaturation. Nucleation is a pre-requisite for crystal growth, which, due to kinetics, is most practically achieved at significantly higher levels of supersaturation than those used for crystal growth. Once the crystal nucleus has formed, crystal growth follows spontaneously (Ataka, 1993; Ducruix and Giege, 1999; McPherson, 1999a).

The crystallization process can be effectively illustrated using a phase diagram. A phase diagram indicates which state (liquid, crystalline or amorphous solid precipitate) is stable under different chemical and physical conditions. While greatly simplified, often only taking into consideration two of the many parameters that can affect solubility and hence crystallization, a phase diagram is of great practical importance in understanding the influence and benefit of manipulating these parameters for crystal optimization. A phase diagram offers a concise view of the thermodynamics not kinetics of different parameters. It provides a means of quantifying the influence of parameters such as the concentrations of protein, precipitant(s), additive(s), etc. on the production of crystals. Thus, phase diagrams provide a practical basis for the design of crystal optimization experiments (Ataka, 1993; Chayen, 2004; Ducruix and Giege, 1999; McPherson, 1999b). Figure 2.1 shows four zones representing different degrees of supersaturation. The precipitation zone of highest supersaturation is where the protein will aggregate and form a precipitate. A labile zone is of moderate supersaturation and represents the conditions where spontaneous,

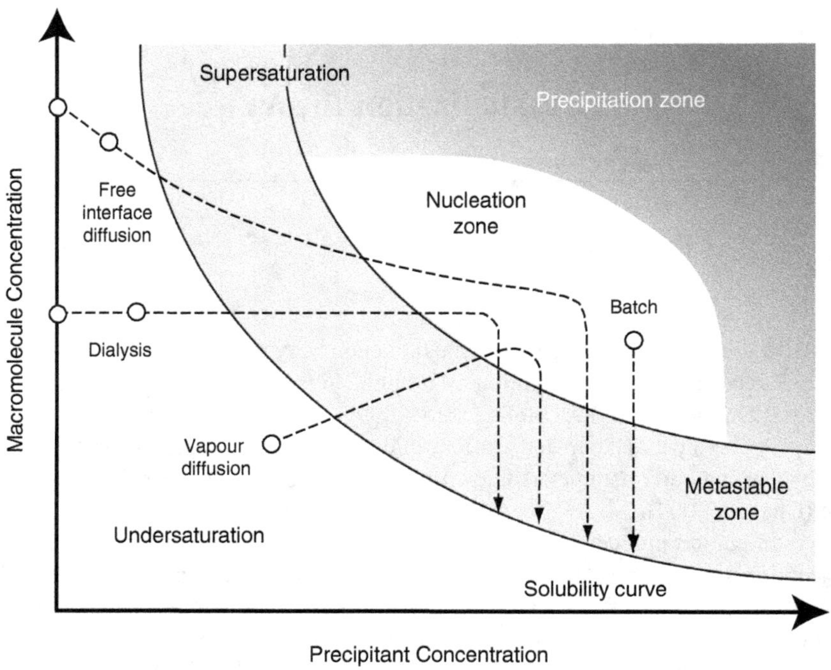

Fig. 2.1. An example of a typical crystallization phase diagram. The figure schematically illustrates four areas: (i) an area of very high supersaturation where the protein will precipitate; (ii) an area of moderate supersaturation where spontaneous nucleation will take place; (iii) an area of lower supersaturation just below the nucleation zone where crystals are stable and may grow but no further nucleation will take place (this area is referred to as the metastable zone that is thought to contain the best conditions for growth of large well-ordered crystals); (iv) an undersaturated area where the protein is fully dissolved and will never crystallize (Chayen (1998) and used with the permission of the IUCr).

homogeneous nucleation takes place. A metastable zone of low supersaturation is where crystal growth from either homogeneous, or heterogeneous (foreign) nuclei takes place, but no additional nucleation occurs. The conditions in this metastable zone are the best conditions for growth of well-ordered crystals. The solubility curve is a dynamic equilibrium where the crystal is stable, and is the endpoint of the typical crystallization experiment. Below the solubility curve is the zone of undersaturation where the protein is in solution and will not crystallize. The undersaturation zone is where most vapour and liquid diffusion crystallization experiments start. This diagram serves as a pictorial description of the process common to many discussions regarding crystallization; however, it must always be kept in mind that the representation is greatly simplified.

Also illustrated in Figure 2.1 are tracks through the crystallization phase diagram that represent different methodologies of crystal growth. In this section, we will cover the methodologies, empirical techniques and direct the reader to some of the detailed literature available on the theoretical studies of macromolecular crystallization.

2.1 Chapter Summary

A basic understanding of crystallization theory exists through the phase-diagram description. More detailed theories are available but no one theory can currently predict crystallization behaviour.

3

Practical methods of crystallization

There are four major techniques for growing crystals of biological macromolecules. These techniques have been in use for over 40 years and include: batch crystallization, vapour diffusion, liquid/liquid diffusion (also known as free interface diffusion) and dialysis. These techniques are schematically illustrated in Figure 3.1. The typical time for macromolecular crystal growth is 1–3 weeks, using any of these techniques; however, the time for obtaining crystals can vary from several hours to as long as a year. Variants of these methods have been developed over the years, some of which are described.

3.1 Batch

Batch crystallization is the oldest and simplest method to employ. The molecule to be crystallized is mixed with a crystallizing agent at the required concentration at the start of the experiment and sealed in an air-tight container, as one variant. Supersaturation is achieved immediately upon mixing. The volume of batch experiments typically range from 50 μl to several millilitres. From a chemical perspective, batch experiments are readily characterized as the starting chemical concentrations are not varied during the course of the experiment.

Microbatch (shown in Figure 3.1(a)) is a technique that is a miniaturized batch method consisting of nanolitre to microlitre experiment drop volumes. Developed in the 1990s, a principal advantage of the microbatch technique is a significant reduction in sample requirements; consuming less sample than classical batch experiments and in the process allowing more sampling of parameter space. In order to prevent evaporation of such small volumes, the crystallization samples are dispensed and incubated under the surface of a low density oil (ρ = 0.83–0.92 g/cm^3), which acts as an inert sealant to protect the drops. A modification of the microbatch technique employs silicone oil or mixtures of paraffin and silicone oils (D'Arcy, 1994; D'Arcy *et al.*, 2004) to envelope the experiment drops. This allows slow and continuous evaporation of the drops, increasing the relative chemical concentrations within the drop, leading to higher levels of supersaturation. In many respects, this method resembles a vapour-diffusion experiment, described below.

Microbatch under oil is compatible with most precipitants, buffers and additives, including detergents. The oils are compatible with the common precipitants such as salts, polyethylene glycols (PEG), jeffamine MPD as well as glycerol and ethanol. Microbatch, should not be used when small volatile organic molecules such as dioxane (an additive used to prevent twinning), phenol, or thymol are present as these chemicals are soluble in the oil (Chayen, 1997).

Classical batch experiments are by design limited in their exploration of the phase diagram. Consequently, multiple trials are required to sample a range of variables such as precipitating agent concentration. In contrast to batch/microbatch, the various diffusion and dialysis methods all reach supersaturation gradually, by attaining an equilibrium between an undersaturated macromolecule solution and a reservoir containing the crystallizing agents, which we now discuss.

3.2 Vapour Diffusion

Vapour diffusion methods, Figure 3.1(b), begin with an aqueous experiment drop containing the macromolecule and the crystallization agents at a concentration lower than that required for the formation of crystals (Bergfors, 1999). This drop (typically from 0.5–10 μl) is equilibrated against a reservoir (0.5–25 ml) containing a solution with crystallization agents at different concentrations than in the experiment drop. While the crystallization cocktail is typically used as the reservoir solution, it is not a requirement that the reservoir solution and crystallization cocktail be the same chemical species (Dunlop and Hazes, 2005; McPherson, 1992; Newman, 2005). Equilibration proceeds by diffusion of the volatile species (typically water until vapour-pressure equilibrium is reached between the experiment drop and the reservoir. As a consequence of the equilibration process, there is an increase in the relative concentration of solutes in the experiment drop. By decreasing the amount of water in the drop, the reservoir acts to gradually concentrate the protein, and other chemicals in the experiment drop. This gradual change and increase in concentration may lead to supersaturation of the protein and subsequent crystallization. This is a passive experiment, as the changes in drop volume and composition are not controlled. This may lead to unanticipated changes in variables such as the pH (Mikol *et al*., 1989). Knowing the pH can be critical as protonation states of ionizable amino acids can have a big impact on the outcome of crystallization trials (McPherson, 1995). Careful notation is critical to ensure the exact experimental protocol can be reproduced if necessary to identify subtle but consequential crystallization variables. Three commonly used vapour-diffusion methods are; hanging drop, where the drop is suspended over a reservoir from the underside of a hydrophobic surface; sitting-drop, where the drop sits in a small concave surface; and sandwich-drop, where the drop makes contact with two narrowly separated surfaces. The sitting and sandwich drop methods can accommodate a larger experiment drop volume than the hanging drop. Larger drop volumes can be used to produce larger crystals (Fox and Karplus, 1993).

3.3 Dialysis

Dialysis – The macromolecule solution is held in a container that is separated from a volume of crystallization cocktail by a membrane with a molecular weight limited permeability, Figure 3.1(c). Equilibration proceeds by diffusion of the small solute molecules (ions, additives, buffers) into the macromolecule solution, whereas the macromolecule remains trapped, unable to pass through the membrane. Unlike vapour-diffusion experiments (where the protein and crystallization cocktail concentrations both increase during equilibration), for the dialysis experiment, the protein concentration is held constant, while the crystallization cocktail concentration increases. The dialysis method is used primarily for optimization once conducive conditions for crystallization have been found using a more efficient method. Currently, the minimum volume of sample solution used in this technique is 5 μl (Chao *et al.*, 2004). At the time of writing this book, the dialysis method has not been automated.

3.4 Free Interface Diffusion

Liquid/liquid diffusion is also known as **free interface diffusion (FID)** (not shown in Figure 3.1 due to its obvious simplicity). In this case equilibration proceeds by the diffusion of two solutions placed in a capillary that creates a concentration gradient. One solution contains the sample (typically 10 μl or more); the other contains the crystallizing agents. The solutions are carefully superimposed and left to slowly mix diffusively, producing a gradient along which precipitate, crystals of various sizes, and clear solution are simultaneously present. This method was least used in the past due to handling difficulties and the requirement for large quantities of sample. In the past 4 years, this method has experienced a revival with the design of new ways of conducting FID trials. Experiments referred to as counter diffusion can now be performed in an X-ray capillary using 1 μl of sample (Garcia-Ruiz *et al.*, 2002; Tanaka *et al.*, 2004). In some cases, the crystals need not be removed from the capillary at any stage, since cryo-protection, soaking with heavy atoms, flash cooling and data collection can in principle all be performed in the capillary (Ng *et al.*, 2003). A crystallization cassette has also been designed for holding many capillaries that can be put into contact with various precipitant solutions, for the use of this method within a high-throughput environment (Ng *et al.*, 2003). A breakthrough in the miniaturization and automation of the FID method has been made with the development by Hansen *et al.*, (2002) who designed a microfluidic chip for rapid screening of protein crystallization conditions. It is interesting to note that until recently, the free interface diffusion method has been used more in microgravity than on earth, because of difficulties in setting up trials due to gravity on earth. The new experimental designs and especially the chip, overcome these difficulties, while at the same time enabling trials to be performed automatically using minimal quantities of sample (Chayen, 2005).

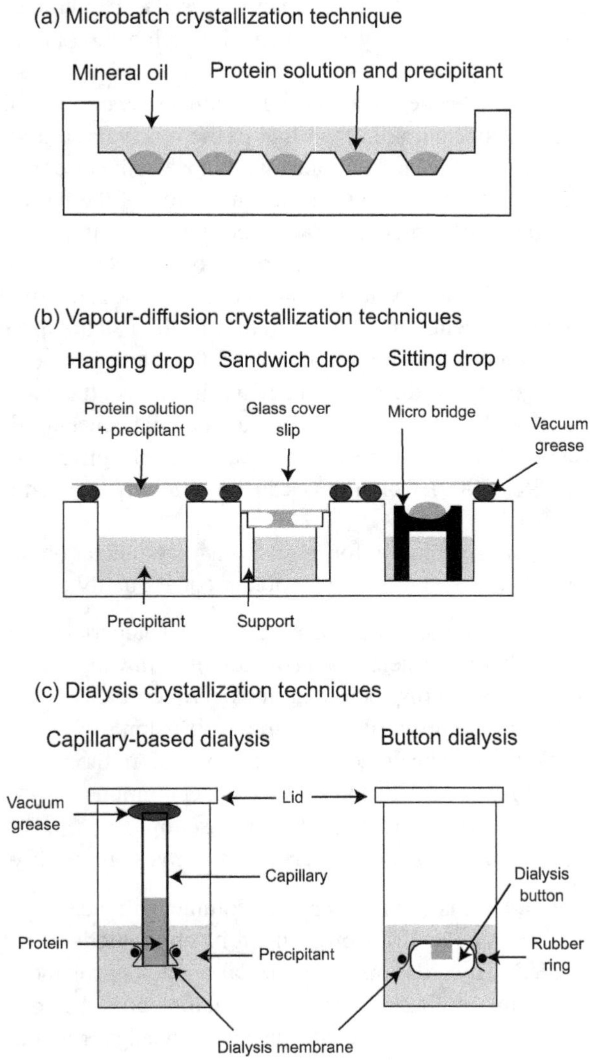

Fig. 3.1. Schematic illustrating the three most common methods of crystallization and variants on those methods. Liquid-liquid diffusion is not illustrated, consisting only of two liquids, precipitant and sample, in physical contact.

3.5 Converting Between Methods

Microbatch and vapour-diffusion methods are each successful in identifying hits during the screening process with a considerable overlap of outcomes (Baldock *et al.*, 1996). However, some hits are unique for each method due to the inherent differences in their kinetics. Moreover, it often happens that a hit obtained using one method cannot be optimized using that method, but conversion to another technique yields sizeable, high-quality crystals. The conversion from one method to another is usually easily achieved by varying the parameters described below (Chayen, 2009).

A vapour-diffusion experiment usually starts with some, or all, of the components in the drop at a lower concentration than that in the reservoir solution. The concentrations of ingredients in the drop equilibrate with those in the reservoir via the vapour phase and are expected to reach the same concentration of the reservoir solution. In reality, the concentration of protein and the other ingredients in the drop at the end of a vapour-diffusion trial, are lower than theoretically expected (Arakali *et al.*, 1995; Fowlis *et al.*, 1988). This must be taken into account when converting from vapour-diffusion conditions to a batch/microbatch setup. In the batch experiment the components are at their final concentration already at the start of the experiment. Thus, when converting from vapour diffusion to batch/microbatch, the concentrations used in the batch setup should be somewhat lower than the final concentrations expected at the end of a vapour-diffusion experiment. Examples of conversions are given in Chayen *et al.,* (1998; 2009) Conditions can be similarly converted from vapour diffusion to microbatch.

There are four rules-of-thumb for converting crystallization conditions from vapour diffusion to batch, or vice versa to obtain equal quality crystals:

1. The final concentration of protein in a batch/microbatch trial is generally 10–20% lower than the starting protein stock solution used in vapour diffusion.
2. The final concentration of precipitant in batch/microbatch is normally 10–20% lower than that in the reservoir of a vapour-diffusion trial.
3. When crystallization takes place very rapidly, before the vapour diffusion has reached equilibrium, significantly lower concentrations (30–40%) of protein and precipitant should be used for the batch setup.
4. The optimal concentrations of buffers and additives are usually held constant.

In the neutron research case, where crystal volume will need to be maximized, conversion from, e.g. vapour diffusion or microbatch methods to batch or dialysis, will likely be needed. Towards this goal special apparatus for moving the crystal-growth conditions within a phase diagram using temperature has been described by Budayova-Spano *et al.* (2007) and methods using knowledge from the phase diagram by Maeda *et al.* (2004).

3.6 Chapter Summary

There are four typical types of crystallization experiment setups with vapour diffusion being the most popular and widely used technique. Each method has many variants. No one crystallization method is superior to another (except when a scale up in volume is needed for neutron diffraction). Each method has its inherent advantages and disadvantages. The choice of method depends on the availability of protein, chemical composition of crystallization trials, ease of set up, and the preferences of the experimenter.

4

Screening

4.1 Screens: Problems and New Developments

Finding initial crystallization conditions for a new macromolecule can be like searching for a needle in a haystack. The first step is to set up screening trials, exposing the macromolecule to numerous different chemical agents in order to find 'hits' or 'leads' that point to conditions that may be conducive to crystallization. Crystals, crystalline precipitate and phase separation are all considered leads that are worth pursuing. Each experimental condition will gradually reveal the solubility phase diagram of the macromolecule. If we refer to the phase diagram discussed in Chapter 2 and plot our results as a function of that phase diagram we begin to see how screening results can be used to determine the conditions that drive the solubility. An example using the batch method is given in Figure 4.1. In this case the conditions have been varied as a function of the macromolecule and precipitant concentration. Because one used the batch method, the initial nucleation condition can be determined but the endpoint is not known. However, one can deduce the regions that the experiment must have traversed from the results. For example, in Figure 4.1(a) no change is seen in the crystallization experiment, the macromolecule is still in an undersaturated state, or has entered the metastable zone without passing through the nucleation or precipitation zones. In Figure 4.1 (b) the result we are looking for, a crystal, has occurred. The experiment started off in the nucleation zone and, based upon the limited nucleation, quickly transitioned into the metastable zone resulting in a few large crystals. If the experiment stayed in the nucleation zone for a longer period of time before transitioning to the metastable zone, more nucleation would take place and large numbers of smaller crystals would result, Figure 4.1(c). If the experiment had an even more extensive stay in the nucleation zone, many tiny crystals would be formed, appearing as (microcrystalline) precipitate, 4.1(d). Precipitate can be amorphous, Figure 4.1(e) and amorphous precipitate can form in the same drop as crystals, Figure 4.1(f) depending on the progression through the phase diagram. This figure is deceptive as many experiments would be required to build up an accurate picture of the phase diagram presented here and many variables can be involved – the phase diagram is not a simple two-dimensional function. One can, however, use it to guide the choice of further conditions to optimize.

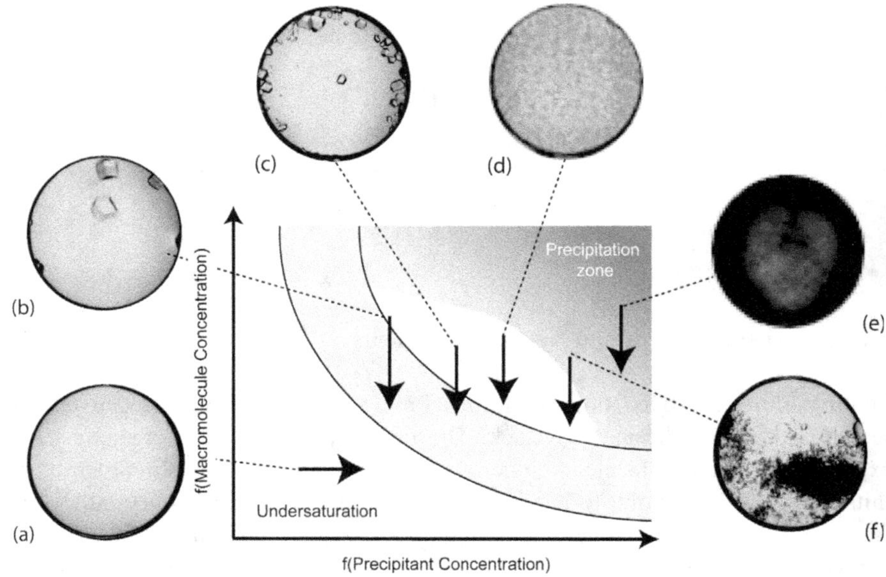

Fig. 4.1. Crystallization screening using the batch method varying macromolecule and precipitant concentration illustrating how the results can be overlaid on a crystallization phase diagram.

During crystallization screening trials, the macromolecule is typically screened against a range of conditions that is pre-dominantly based on those that have been successful in the past. The first commercial screen was one based on a compilation of crystallization results (Jancarik and Kim, 1991) and was termed a sparse-matrix screen from its intended unbiased attempts to sample the multidimensional space of crystallization parameters. Carter and co-workers (1997) designed incomplete factorial screens that made a major advance in the application of partial factorial designs to the area of crystal growth. In these designs, relative levels of important chemical factors are sampled to achieve good coverage and good balance in the sampling. Additional screens have surfaced, based either on precipitants and additives the usefulness of which was discovered later (Cudney *et al.*, 1994b) or based on the much wider record of crystal producing (positive) results provided by the Biological Macromolecule Crystallization Database (Gilliland *et al.*, 1996; Gilliland *et al.*, 2002). The incomplete factorial screens are more efficient at hitting on initial crystallization conditions than systematic grid screens (Segelke, 2001), the latter having the advantage of providing useful solubility information even before crystals are obtained (Brzozowski and Walton, 2001; Haire, 1999; Ries-Kautt, 1999). While the latter are frequently used as a second-tier approach, there is no consensus on a single approach, since it is claimed that properly statistically balanced incomplete factorial screens can be more effective even as a first choice (Delucas *et al.*, 2005a). Other sampling strategies based on orthogonal arrays (Kingston *et al.*, 1994) and random samplings (Shieh *et al.*, 1995) have been described and used to great effect.

Apart from not providing systematic information on the solubility behaviour of the sample, an additional problem of pre-formed sparse-matrix screens is that the statistics become increasingly biased, i.e. accumulating results based on repeated sampling of the same chemical space. As new trials concentrate around conditions that have already been selected to be part of the screen, new positive results tend to cluster around the old ones and other possible conditions remain undiscovered. Thus, a sort of vicious circle is established (though not completely vicious, or it would have been abandoned) as literature reports only the successful crystallization conditions for proteins, the majority of which are derived from a common set of commercial cocktails. Thus, careful data mining and assessment of negative results is necessary (Page and Stevens, 2004; Rupp and Wang, 2004) to survey, evaluate, and rationally expand the chemical space sampled in crystallization cocktails. A systematic way to somewhat reduce the bias would be to periodically reassess the screens in view of new published results, awarding a much higher 'score' to rare conditions that have given 'hits', than to ones that are well represented in standard screens.

Neural network methodology, initially used in crystallography for predicting crystal growth (Noever *et al.*, 1996) has been applied to the prediction of crystallization conditions (DeLucas *et al.*, 2005b). Relying on results from an incomplete factorial screen, an aim is to train the network to predict the outcomes of non-performed (virtual) trials, i.e. of the full factorial screen. To date, this strategy has not been widely adopted, perhaps due to the limited data stets available for training of the neural network.

Another approach has been to use parameters that can be calculated from amino acid sequence information with results of previous crystallizations. This has particular relevance to structural genomics projects where failure as well as success is routinely noted. Slabinski *et al.*, (2007a) have used this data to determine several features of macromolecules that correlate strongly with successful production and crystallization. These features are used by a webserver called XtalPred (Slabinski *et al.*, 2007b) to calculate a crystallization class ranging from optimal to very difficult. The features used include length, isoelectric point, the grand average of hydropathy (known as the 'gravy index' (Kyte and Doolittle, 1982)), possible disordered regions, an instability index, and the percentage of coiled structure. It is possible to perform virtual truncations to examine the effect of these on the potential for crystallization. However, this method can be biased on the population used to build up the data. In several cases macromolecules that are readily crystallized experimentally are classified by the algorithms as very difficult to crystallize. It is probable that this approach will be successful for related samples but not as a generally applicable technique until a much larger database of results is established.

In the Hauptman-Woodward Medical Research Institute High-Throughput Screening Laboratory our initial screen is based on an incomplete factorial sampling of crystallization space. We classify the results and then present them in chemical space (Snell *et al.*, 2008) that provides a vector for optimization of the crystallization conditions. Examples of this are illustrated in Figures 4.2 and 4.3. In Figure 4.2 a small portion of the incomplete factorial screen is shown with the results coded

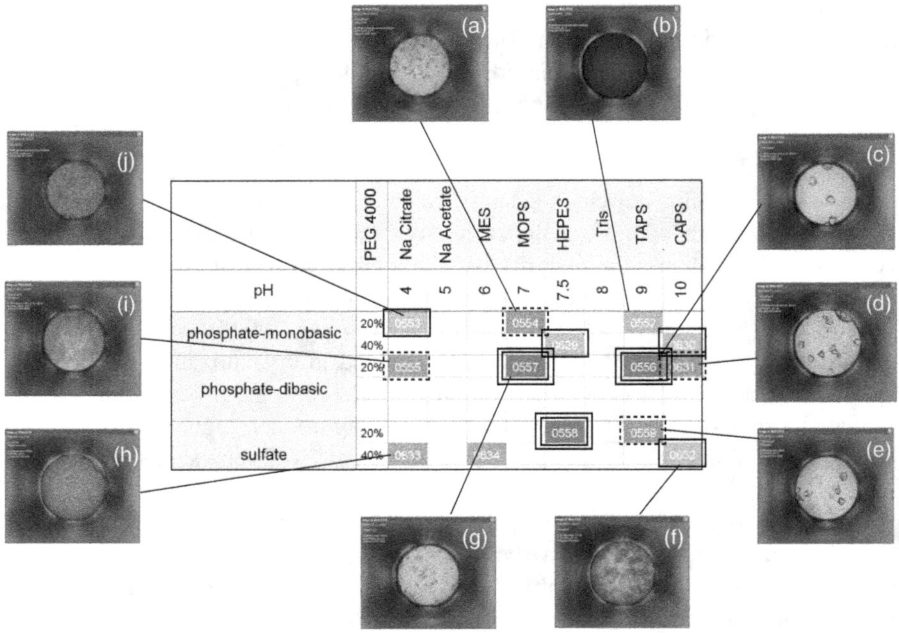

Fig. 4.2. A chemical space map where the cells are normally coloured according to the crystallization result. The reader is referred to Snell *et al.* (2008) for the original figure in colour. Here, crystals are shown with a double box around the cell, precipitate with a single box. Images that show a combination of results are indicated with a dashed box. Other classifications are not highlighted. Images (a), (d), (e) and (f) have been classified as a mixture of crystals and precipitate. Image (b) has been classified as clear with a dark colour but no apparent texture. In the context of its chemically related neighbours, this classification is probably incorrect and the image should be reclassified as precipitate. Images (c) and (g) are crystals only, and finally image (h) shows evidence of a skin and precipitate. The uncoloured blocks indicate areas of the incomplete factorial screen that are unsampled. Used with the permission of the authors and the IUCr.

according to their classification and then presented in the chemical space using a software called AutoSherlock (Nagel *et al.*, 2008). At low pH there is a crystalline precipitate. As pH increases, small crystals occur that translate to larger crystals at pH 9 for both the phosphate-dibasic and sulphate salts but not the phosphate-monobasic. The white areas in the colour-coded table indicate regions that are not sampled experimentally. In this case, from our knowledge of the crystallization phase diagram, we can predict that pH is an influential variable affecting solubility. The single variable of pH manipulates macromolecular solubility over the entire range of outcomes from precipitation to the undersaturated region passing through areas of interest for crystallization. In Figure 4.3(a) the same analysis was used and the results presented in a grid as a function of pH and PEG molecular weight. In this case there is a clear vector showing a direction for optimization of crystallization conditions,

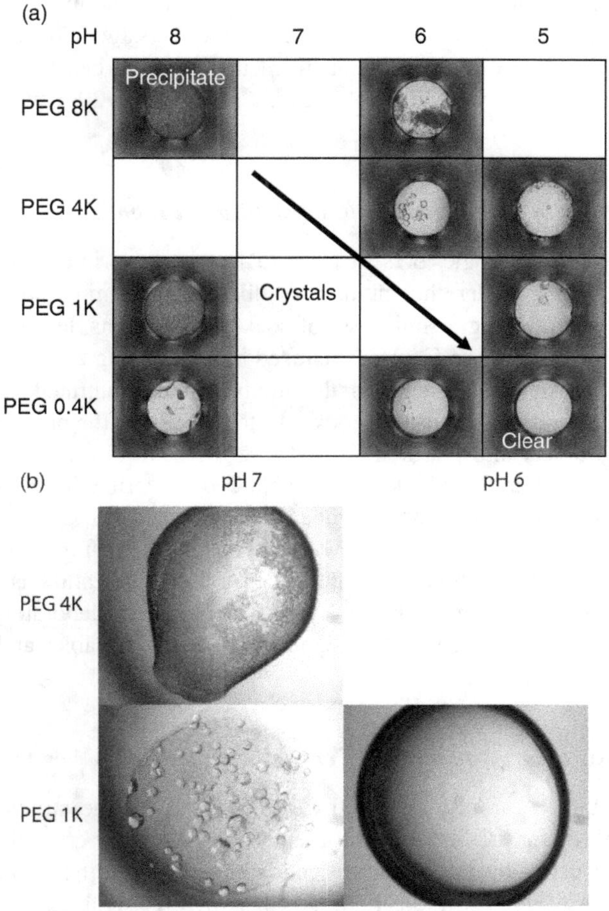

Fig. 4.3. (a) Replotting chemical space data as a function of pH and PEG type to show the dependence of the results on these two variables (b) Scaled-up crystallization trials for promising conditions not sampled by the incomplete factorial matrix (Snell *et al.*, 2008), used with the permission of the authors and the IUCr.

highlighting potentially important conditions that were not sampled during the initial screen. In Figure 4.3(b) the relevant unsampled conditions have now been sampled in a scaled-up experiment. The optimum conditions were identified by analysis of the initial screen, and the trends were confirmed by the scale-up.

4.2 Automation and Miniaturization of Screening Procedures

The past several years have seen some great efficiencies made in the field of macromolecular crystallization. The ability to dispense trials at nanolitre volumes in a high-throughput mode has cut the time of setting up experiments from weeks to minutes, a scenario that was unimaginable just a few years ago (Chayen, 2004). Even more incredible, is the revelation that diffracting crystals can be produced from crystallization experiments with volumes as small as 5–20 nanolitres (Hansen *et al.*, 2002).

4.2.1 The effect of high-throughput: crystallization robotics

High-throughput technologies present a way to do more experiments, faster, and with less material. They render the initial crystallization screening process, an often repetitive process utilizing a similar set of cocktail conditions, less labour intensive and sample consuming. The former is realized by automating as many stages of the procedure as possible, starting with the mixing of concentrated chemical stock solutions into a set of crystallization cocktails, to dispensing the protein plus precipitant mixture droplets into plates (Bard *et al.*, 2004; Luft *et al.*, 2003), sealing the plates, monitoring (Bard *et al.*, 2004; Luft *et al.*, 2003), scoring the outcomes with the use of image recognition software (Cumbaa and Jurisica, 2005; Cumbaa *et al.*, 2003; Wilson, 2002) and even suggesting new experiments based on those outcomes. The full power of high-throughput technologies to aid crystallization is only recently being realized with the combination of results' tracking and laboratory information management systems (LIMS) that can correlate the many variables and stages of the crystallization process.

4.2.2 Analysis of large quantities of crystallization data

The high-throughput procedures, used to great effect by the structural genomics community, allow for the first time intra- and inter-laboratory crystallization studies that utilize both crystallization successes and failures, along with an accounting of the total number of trials. The indications are that there is significant redundancy of conditions between and within commercially available crystallization screens; and minimal screens for crystallization hit identification can be developed. For example, at the Joint Centre for Structural Genomics, Scripps Research Institute, it was found that 67 out of 480 conditions tried would have been sufficient to crystallize 84% of the total 465 proteins that were crystallized, whilst an additional 29 conditions would have raised that percentage to 98% (Page and Stevens, 2004). Rupp and colleagues estimate that <300 trials (i.e. three 96-well plates) should be sufficient to find crystallization conditions, if these trials fail, time and resources would best be spent in changing organism or construct, or looking for genetic orthologues (Chayen and Saridakis, 2008; Rupp and Wang, 2004).

Other high-throughput laboratories use larger screens. For example the one in use at the Hauptman-Woodward Medical Research Institute High-Throughput Screening (HTS) Laboratory (Luft *et al.*, 2001) described above uses 1536 different conditions. The philosophy here is somewhat different. While looking for crystallization leads the incomplete factorial sampling of a large area of chemical space allows the

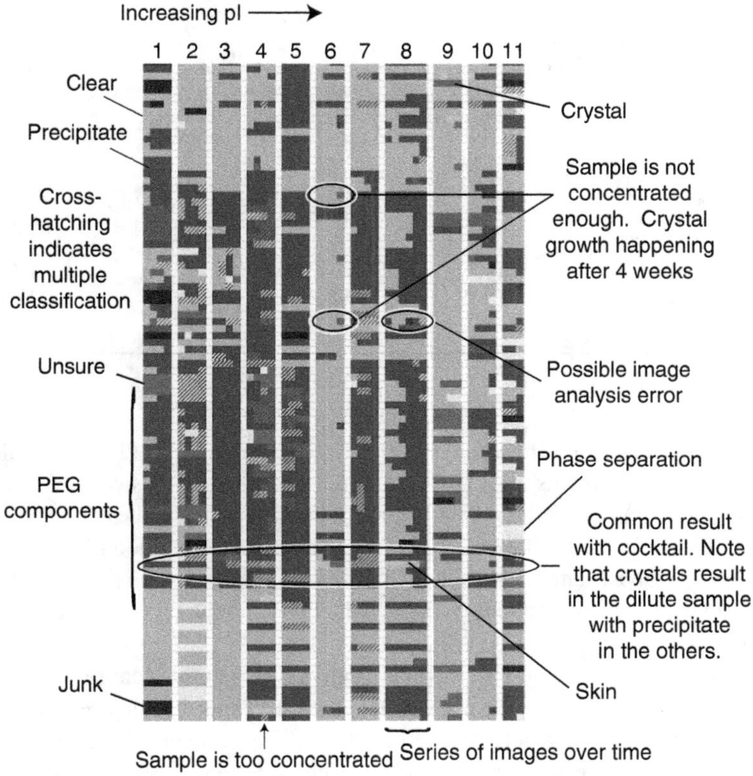

Fig. 4.4. 'Ribbon' cocktail response diagram showing 11 different samples (vertical) against 90/1536 different crystallization cocktails and time – a fingerprint for each sample (horizontal). In this case colour information has been converted to greyscale but trends are still visible (Grant, Luft and Snell, unpublished).

crystallization phase diagram to be mapped and also provides a more complete picture of the chemical space (Snell *et al.*, 2008). While requiring more sample volume, this provides three advantages over screening with a smaller number of conditions; a fingerprint of the sample behaviour under different biochemical conditions, the identification of multiple hits for optimization and a direction to take with that optimization. We have discussed the use of this screening to provide a direction for optimization above. The concept of chemically fingerprinting a sample comes into play for looking for patterns in crystallization. In Figure 4.4 we have plotted a small number of screening results for 11 different macromolecular samples. These results have been colour coded according to the outcome. Patterns can be seen telling us simple things such as the concentration may be too low or too high. More complex pictures can begin to emerge; a number of PEG conditions strongly precipitate 7 out of 11 macromolecules, and one cocktail causes precipitation in almost all cases, outcomes that have potentially been classified incorrectly are highlighted. Figure 4.4 illustrates the

Table 4.1. Summary of research progress from worldwide structural genomics centres up to March 2009 from data at Target DB (Chen *et al.*, 2004)

Status	Total number of targets	% relative to cloned	% relative to expressed	% relative to purified	% relative to crystallized
Cloned	146 779	100.0	-	-	-
Expressed	99 594	67.9	100.0	-	-
Soluble	37 750	25.7	37.9	-	-
Purified	34 993	24.5	36.1	100.0	-
Crystallized	12 616	8.6	12.7	35.1	100.0
Diffraction	5753	3.9	5.8	16.0	45.6
Structure	4590	3.1	4.6	12.8	36.4

potential of this analysis, showing only 90/1536 conditions for 11 macromolecules. The HTS database contains over 10 000 different macromolecules with 1536 cocktail conditions and images of the crystallization result for each of these over at least 4 weeks duration. The size of this database is currently over 90 million images but using it will require robust automated image analysis techniques as it is impossible for a person, or even team of people to cope with this amount of data in a reasonable time frame.

Table 4.1 shows the sum of results from 17 structural genomics/proteomics projects worldwide through March 2009, demonstrating the success rate going from cloned protein to structure determination (The numbers for each individual project vary as they are dealing with different genomes and have started at different times). The results demonstrate that out of 146 779 samples successfully cloned, 68% of these were expressed, 26% were soluble and 25% were purified. Of these soluble purified samples, 5753 well-diffracting crystals were produced (4% of the originally cloned samples) leading to the determination of 4590 structures at that time. It transpires that over half of the trials that were set up from pure soluble proteins had produced crystals, but only about half of those crystals were diffracting ones. It is evident that a large failure rate in the structural genomics process (and by analogy crystallography itself) is in the step between making purified protein and obtaining diffraction-quality crystals. Even soluble and highly pure macromolecules are hard to crystallize. Some proteins (referred to as 'low hanging fruit') do crystallize 'easily' during the initial screening stage but many trials are yielding microcrystals or poorly-ordered ones. In spite of the impressive advances in automation, and now generating approximately half the structures deposited per year, high throughput has not led to high output, clearly demonstrating that crystallization is a fundamental limiting factor in the field. The problem is how to convert the >80% of the soluble, purified samples that remain structurally uncharacterized into useful analysable crystals. It is evident that screening procedures are not sufficient to produce the expected output and further effort is required to find crystallization strategies that will deliver the desired results (Chayen, 2003).

4.2.3 Experiment volume considerations

High-throughput automated techniques can set up many samples quickly. They are also highly accurate and reproducible such that small quantities of sample can be used. However, miniaturization can both increase the number of 'false negatives' i.e. conditions that would have given crystalline material if tried at larger volumes, and, lead to positive results that are not easily scaled up from the nanolitre scale (Carter *et al.*, 2005). Drop volumes in the hundreds of nanolitres range can therefore have some advantages over those in the tens of nanolitres. Several laboratories have demonstrated that using 100 nl of protein solution + 100 nl of crystallizing agents will produce crystals that can be X-rayed directly from such drops without the need to scale up the volume (e.g. Walter *et al.*, 2003). For individual laboratories who crystallize a limited number of proteins it is sufficient to conduct trials either manually or by using automated hand-held pipettes (Chayen and Saridakis, 2008).

4.2.4 Imaging and monitoring of crystallization trials

Dispensing trials with such speed and ease has shifted the bottleneck to observation and monitoring and analysis of the vast numbers of trials. Major efforts are underway to design image-processing equipment for automated follow-up and analysis of the results. Many of these efforts emphasize the identification of several specific categories of outcomes related to crystallization leads (Bern *et al.*, 2004; Berry *et al.*, 2006; Cumbaa and Jurisica, 2005; Cumbaa *et al.*, 2003; Kawabata *et al.*, 2006; Mayo *et al.*, 2005; Miyatake *et al.*, 2005; Walker *et al.*, 2007; Wilson, 2002; Wilson and Main, 2000; Zuk and Ward, 1991). Full automation of the visualization and monitoring of trials is a long-sought after goal.

4.3 Chapter Summary

Although our theoretical understanding of macromolecular crystallization is incomplete, an empirical approach can be used to yield results that can be interpreted and acted upon to optimize the process. Having described this initial screening, the first step of the crystallization process, the next chapter will deal with optimization, i.e. how to turn the leads obtained during screening into high-quality crystals.

5

Optimization

5.1 Practical Uses of the Crystallization Phase Diagram

It is important to understand that not all macromolecules are certain to crystallize. There are many examples of macromolecules that appear to be natively unfolded. For those that will crystallize we can use the phase diagram described in Chapter 2, to empirically understand and predict their crystallization behaviour.

In an ideal experiment, once nuclei have formed, the concentration of protein in solution will decrease as protein is sequestered by the crystalline solid phase, thereby naturally leading the system into the metastable zone where growth should occur, without the formation of additional nuclei (Bergfors, 1999; Ducruix and Giege, 1999; McPherson, 1999a). However, this ideal depiction occurs infrequently. More often than not, excess nucleation occurs, resulting in the formation of many small crystals. The initial aim is to devise methods that will enable the experimenter to lead the experiment from nucleation to growth, and to produce only a few crystals of sufficient size for diffraction analysis. The diffraction analysis will then guide further optimization.

The phase-diagram concept, despite its simplicity, provides basic and important knowledge to rationally devise optimization experiments for macromolecular crystallization. While it may not be possible to distinguish the location of a clear drop relative to the phase diagram as an isolated data point, it is possible to determine the proximity of a clear drop to the metastable zone empirically by reference to the surrounding experiments when they are presented in chemical space (Snell *et al.*, 2008). For example, clear drops located near a precipitation boundary, and near or next to crystals along a chemically sensible vector, are more likely to be located in a metastable region than those bounded by other clear drops. Similarly, precipitation indicates that the solution is significantly more supersaturated than required for crystallization; the presence of crystals indicates that the experiment has entered, or passed through the labile region. By arranging crystallization screening results for a particular macromolecule in chemical space, we can globally portray the complexity and sensitivity of the phase diagram to specific chemical variables. In this manner, we do not observe isolated data, but rather observe outcomes in the information-rich context of a phase diagram.

Quantitative phase diagrams can provide an accurate picture of crystallization space, but they require numerous solubility measurements. The solubility is defined as the concentration of the macromolecule in solution where dissolution and deposition of molecules on the crystal surface are in equilibrium. Reaching equilibrium can take several months (Ataka, 1993 and references therein) because macromolecules diffuse slowly. An additional limiting factor is that solubility measurements require a large amount of sample. A method for pinpointing the metastable zone has been devised based on the second virial coefficient, which is a measure of pair interaction between particles (Wilson, 2003), but the procedure to date is complicated for immediate application to routine use. The area of conditions called the 'metastable zone' is situated between the solubility and supersolubility curves on the crystallization phase diagram. The supersolubility curve is defined as the line that separates the conditions where spontaneous nucleation (or phase separation or precipitation) occurs, from those where the crystallization solution remains clear if no nucleation event occurs (Ducruix and Giege, 1999).

The supersolubility curve is experimentally much easier and quicker to determine than the solubility curve, and in practice, provides similar information. To construct it, one must set up crystallization trials, varying at least two conditions (one of which is typically the protein concentration) and plot their outcomes on a two-dimensional parameter grid (Chayen and Saridakis, 2008). The supersolubility curve can be obtained rapidly using robots and can be used to separate conditions of nucleation (labile) and growth (metastable) for optimization using seeding, nucleants (Saridakis and Chayen 2009) and other means. A diagram containing the fully detailed supersolubility curve (and not the solubility curve) is called a 'working phase diagram' (Saridakis and Chayen, 2003).

The first 'working phase diagram' which was generated for the purpose of seeding was reported by Khorkhin *et al.* (1995). They constructed such a phase diagram in approximately 4 hours using the automated microbatch technique to dispense six protein concentration values versus 24 different precipitant concentrations. Conditions for microseeding were predicted from that phase diagram and subsequent to seeding, crystals of alcohol dehydrogenase diffracting to 2 Å were grown reproducibly in 2 μl drops. More recently, similar phase diagrams were constructed automatically in order to find conditions for successfully inserting external nucleants (Chayen *et al.*, 2006). Quantitative phase diagrams are not typically employed unless there are specific experiments that are aided by knowledge of the crystallization space, e.g. when growing large crystals for neutron diffraction (Maeda *et al.*, 2004).

Crystallization is a dynamic process. This is shown in Figure 5.1 where an initial 'hit' has been optimized by varying the ratio of the protein to precipitant (horizontal axis) and the temperature (vertical axis). The results from this drop volume ratio/temperature (DVR/T) optimization technique (Luft *et al.*, 2007) are shown over four weeks. The initial hit is the central condition that shows signs of action (observable phase changes) at the first week that eventually develops into a crystal. Interestingly, several other phase transitions are seen, crystals appear in precipitated experiments and what initially appeared to be skin formation forms crystals on its surface. In crystallization studies, the only experiment that fails to provide solubility data are

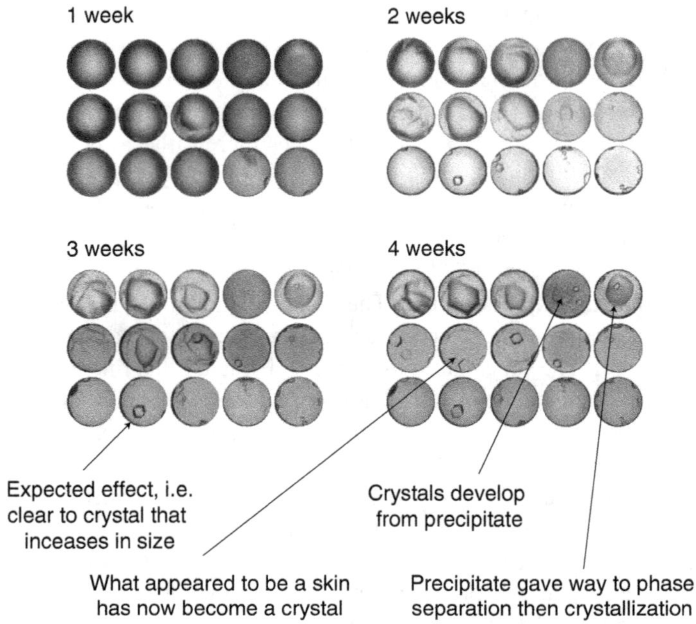

Fig. 5.1. Optimization of crystallization conditions using DVR/T methods (Luft *et al.*, 2007) showing outcomes over time. The expected process of crystallization exemplified by the phase diagram occurs in the bottom row. Over time, clear experiments become crystalline. There are other results not explained well by the typical phase diagram, e.g. a macromolecular skin on the drop leading to crystals or the formation of large crystals from crystalline precipitate (Ostwald ripening).

experiments that are not set up. Each experiment provides another piece of solubility data. When the data is considered as part of a composite, rather than an individual component, it creates more expansive and greater detailed versions of the protein's phase diagram.

In the example of Figure 5.1 the crystals that result from precipitate (possibly from Ostwald ripening (Ostwald, 1896) described by Ng *et al.* (1996)) and those that occur from a skin are not well explained by the schematic phase diagram in Figure 2.1. A closer look at crystallization phase diagrams and the information they contain is provided by Asherie (2004).

5.2 Methods for Separating Nucleation and Growth

In an ideal experiment, once nuclei have formed, the solution concentration of protein will decrease. This naturally leads the system into the metastable zone where crystal growth should occur without the formation of additional nuclei (Bergfors, 1999; Ducruix and Giege, 1999; McPherson, 1999b). As mentioned above, one wants to guide the crystallization system to form one or very few nuclei that will turn into single crystals, ideally of good size and quality. A variety of methods to achieve this are described in the next sections.

5.2.1 Seeding

The most common means of separating nucleation and growth is by seeding, which involves inserting crystals, or crystalline material directly into the metastable zone (e.g. Bergfors, 2003; D'Arcy *et al.*, 2004; Stura, 1999).

There are three types of seeding:

1. macroseeding where single crystals are transferred;
2. microseeding in which submicroscopic seeds, too small to be distinguished individually are transferred;
3. streak seeding where tiny clusters of crystals, precipitate, spherulites and even oils are picked up with a whisker (typically from a cat), or hair from a horse's tail, and transferred into a crystallization drop at lower supersaturation.

Of the three methods, macroseeding is the most technically difficult. It is labour intensive, as it requires deftness and care not to break the parent crystal during repetitive transfer of the crystal through several washes, and requires proper solution conditions to ensure the crystal does not dissolve, or cause immediate precipitation of the drop into which it is being transferred for growth. In most cases, macroseeding is not required, and microseeding is quite effective.

Microseeds can be prepared and introduced into an experiment drop in a variety of ways. The crystals can be pulverized (smashed) into crystalline particles by tissue homogenizers, sonication, vortexing, seed beads, glass rods, or other utensils (Bergfors, 2003). Luft and DeTitta (1999) discuss the advantages and disadvantages of different pulverization methods. The seed stock is taken through a series of dilutions in a mother liquor and thus will generate mixtures with different concentrations of nuclei. The dilution with the optimum number of nuclei is not known a priori and needs to be determined experimentally. Experience has shown that dilutions between 10^{-2} and 10^{-7} have mostly worked well. The nuclei can be pipetted from the stock as a small aliquot or transferred with a seeding wand, which is dipped into the microseed mixture to pick up seeds and then touched, stirred, or streaked across the surface of the new drop. When microseeding, it can be difficult to control the number of seeds that are transferred (Bergfors, 2003) unless serial dilution techniques are used.

Streak seeding (Stura and Wilson, 1991) is the easiest and quickest microseeding method. An animal whisker is inserted into the drop containing crystals, precipitate, etc. and is then drawn through a freshly prepared experiment drop, depositing the seeds along the path taken by the whisker through the drop, the streak line. Crystals then grow along the streak line. Seeding wands can be made from thin glass rods, platinum wire and acupuncture needles, but animal whiskers or hairs have an advantage due to the overlapping cuticles that are able to capture the seeds. It is interesting to note that streak seeding can also be conducted in microbatch under oil, where the oil does not wipe away the seeds, as might be expected.

Recently, seeding has been automated and also implemented in screening experiments. A new automated method to tackle cases where spontaneous nucleation was low or the crystal morphology was poor was devised by D'Arcy *et al.*, (2007). This method achieves both screening and optimization by incorporating the addition of seeds into the screening procedure using a standard crystallization robot. In practice, crystals grown in one set of conditions are seeded into a secondary screen of 96 crystallization solutions. The seed stocks can be stored at 193 K and can survive many cycles of freezing and thawing without a decrease in the nucleation effect observed. In all cases tested, an improvement in the number of hits was observed, which varied from a 2.7-fold to a 65-fold increase compared to the initial screens. In some cases, the nucleation could be controlled and the crystal morphology considerably improved by simply using a more diluted seed stock (D'Arcy *et al.*, 2007).

Although seeding is very successful and it is the most widely used method for separation of nucleation and growth (Stura, 1999 and refs therein), seeding often involves handling of fragile crystals, and other than streaking, seeding manoeuvres are not very simple to perform. Insertion of external nucleants is invasive and can give rise to disturbance in the trial drops. A non-invasive simple means of separating nucleation and growth by changing the temperature at which trials are incubated has been successfully applied to several proteins (e.g. Haire Lloyd, 1996; Rosenberger *et al.*, 1993). Another approach is to initiate the crystallization process at conditions that induce nucleation and subsequently lead the system to metastable conditions that promote optimal growth.

5.2.2 Dilution techniques

An alternative means of conducting trials with a similar outcome to seeding, but without the need to handle crystals or nucleants, is by dilution of experiment drops after spontaneous nucleation occurs, effectively decreasing the level of supersaturation to limit nucleation, while driving the system to a metastable state. An additional benefit of dilution methods is that they are more amenable to automation compared with seeding techniques.

A very common occurrence in crystallization is the formation of clusters of non-diffracting crystals that cannot be improved by merely changing the crystallization parameters. In such cases a working phase diagram that provides the supersolubility curve can be generated (either manually or using a robot).

In the case of hanging drops, the coverslips holding the drops are incubated for some time over reservoir solutions that normally give many small crystals. After a given time (before crystals are visible), the coverslips are transferred over reservoirs containing conditions that would normally yield clear drops. This results in dilution of the drops as they are now over reservoirs with lower precipitant concentrations. The time of transfer is selected by reference to the time in which it took to see the first crystals in the initial screens. For example, if crystals appeared within 24 h, nucleation

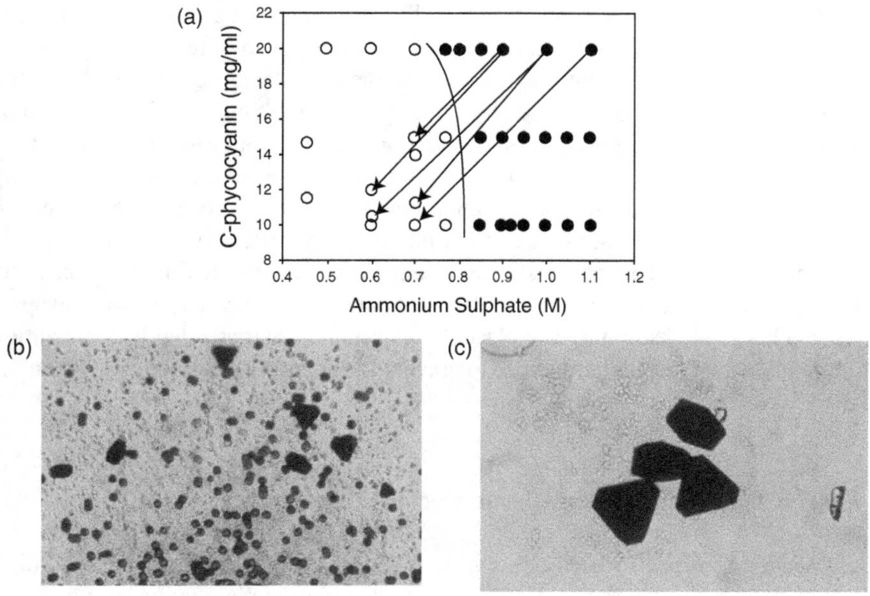

Fig. 5.2. Crystallization of c-phycocyanin by dynamic separation of nucleation and growth. (a) Supersolubility curve of the phase diagram of c-phycocyanin. The arrows correspond to transfer from spontaneous nucleation conditions (black circles) to metastable conditions (open circles). (b) Typical crystals grown under spontaneous-nucleation conditions. (c) Crystals diffracting to 1.45 Å obtained by the transfers shown in (a). Size of largest crystal: $0.15 \times 0.25 \times 0.25$ mm. The photographs were taken at the same magnification. Figure taken from Chayen (2004) used with permission from Elsevier Press.

would have occurred at any time between setting-up the experiments to several hours before the crystals appear. In this case, the transfers should be done at intervals of 1–2 h after set up. Trials that are transferred too soon will produce clear drops, while those that are transferred too late will over-nucleate and produce many small crystals. This technique has produced significant improvement in crystal diffraction of a number of proteins (e.g. Krengel *et al.*, 2006; Saridakis and Chayen, 2003).

Figure 5.2 shows crystals of c-phycocyanin obtained by such a transfer experiment. The crystals reproducibly diffracted to 1.45 Å compared to 1.65 Å using standard optimization techniques. The crystal structure has been solved by Nield *et al.* (2003).

Dilution, as well as the induction and arrest of nucleation can also be achieved in microbatch trials. Using microbatch it is easier to automate such experiments. Saridakis *et al.* (1994) employed the automated microbatch technique to establish a working phase diagram for the enzyme carboxypeptidase G_2. The concentrations of the protein and precipitant were varied, while pH and temperature were kept constant. The conditions for nucleation (i.e. conditions that would produce crystals if the trials were left undisturbed) were found and 2-μl drops were set up as microbatch trials under these conditions. At various time intervals after set up of the experiments, the robot was programmed to automatically insert the dispensing tip into the drops and add buffer or protein solution, thereby diluting the trials. Single diffracting crystals were routinely attained equivalent to the best, very rarely obtained without using the dilution procedure. It is apparent that the dilution prevented excess nucleation by backing off from the nucleation conditions at the time when just one or two nuclei had formed, thereby facilitating the growth of one or a few single crystals.

Induction and subsequent arrest of nucleation can be performed in microbatch by setting up the trials under a very thin layer of paraffin oil, thereby allowing evaporation and hence nucleation to take place. Evaporation is later arrested by the addition of more oil to the crystallization plate to prevent further evaporation of water from the trials (Chayen and Saridakis, 2002).

5.3 The Application of Light-Scattering Techniques

Light scattering can also be used as a diagnostic in crystallization. Wilson (2003) has reviewed both static and dynamic light scattering techniques to monitor pre-nucleation and post-nucleation growth.

5.3.1 Static light scattering

Static light scattering, also referred to as classical or Rayleigh light scattering, measures the average intensity of light scattered by a protein solution of defined concentration in excess of that scattered by background sources (solvent, crystallizing agents, stray light, etc.). It yields the weight-average molecular weight of the solution contents, possibly impure. The working equation appropriate for most protein molecules is:

$$\frac{Kc}{R_{90}} = \frac{1}{M_{\mathrm{W}}} + 2Bc + \ldots \tag{5.1}$$

where K is an optical constant given by:

$$K = \frac{4\pi^2 n^2 (\mathrm{d}n/\mathrm{d}c)^2}{N_{\mathrm{A}}\lambda^4} \tag{5.2}$$

and n is the refractive index of the solvent, dn/dc is the refractive index increment of the solution, N_A is Avogadro's number (per mole), λ is the wavelength of the vertically polarized incident light in vacuum, R_{90} is the excess Rayleigh factor at a scattering angle of 90 degrees, c is the concentration of the protein, M_w is the weight-average molecular weight of the protein and B is the 2nd virial coefficient. A useful further parameter to define is the supersaturation of a solution, σ, which is c/s, where c is defined above and s is the solubility. Note that, unlike small-angle X-ray scattering, the radius of gyration, R_G, is not normally determined by static light scattering since most biological macromolecules are smaller than $\lambda/20$ ($\sim$30 nm), and there is no measurable angular dependence of the scattered light intensity, a requirement for reliable R_G measurement.

It is established both experimentally and theoretically (see references in Wilson (2003)) that a range of second virial coefficient values, in eqn (5.1), is correlated with solution conditions that are favorable to crystallization. This so-called 'crystallization slot,' corresponding to B_{22} values in the range of about -1×10^{-4} to -8×10^{-4} and indicates macromolecule–macromolecule interactions that are slightly to moderately attractive. This crystallization slot is an effective guide to direct changes in particular solution parameters (pH, temperature, crystallization agent concentration, etc.) that will increase the probability of a successful crystallization trial.

5.3.2 Dynamic light scattering

Dynamic light scattering (also known as quasi-elastic light scattering – QUELS) was initially used for predictions indicating which biological systems will crystallize and which will not (Ferre-D'Amare and Burley, 1994; Kam *et al.*, 1978; Malkin and McPherson, 1993; Mikol *et al.*, 1990; Veesler *et al.*, 1994; Zulauf and Darcy, 1992). It is particularly sensitive for the detection of macromolecular aggregates as the scattered intensity is proportional to the square of the molecular weight of the scattering molecules. Zulauf and Darcy (1992) found an empirical method to screen for proteins that had the best chance of crystallizing by examining protein solutions at low concentration and in the absence of any precipitating agents using QUELS. The authors, and later Ferre-D'Amare and Burley (1994) found that proteins showing a tendency to form distributions of aggregates do not crystallize in the majority of cases (approx. 80%, (McPherson *et al.*, 1995)). In a number of cases these aggregates can simply be removed by filtration, centrifugation or a change of the buffer conditions. Otherwise, further purification or modifications may be necessary before investing time and effort in attempting crystallization (Zulauf and Darcy, 1992). This method is now widely applied (Borgstahl, 2007) due to its ease of application, rapid measurement and analysis, and small, recoverable sample requirements.

Dynamic light scattering can also be used for optimization. One of the most effective times to interact with a crystallization experiment is soon after the formation of the first critical size nuclei that will eventually form the crystal. By the time nuclei form or crystals can be observed under the microscope, they have already reached a size of approximately 5 μm and many nuclei are likely to have formed (Chayen and Saridakis, 2002). Dynamic light scattering offers a size resolution of particles in optically transparent aqueous samples some three orders of magnitude of resolution scale below an optical microscope and is therefore a useful tool for an early, non-invasive, *in-situ* observation of a crystallization event, before it becomes visible with a light microscope. In practice, a laser is focused onto the protein solution as the saturation and nucleation processes are occurring and the light scattered by the particles within the solution, such as protein molecules or aggregates, is collected.

The autocorrelation function of the scattered light intensity is analysed and gives the diffusion coefficients and hence the hydrodynamic radii of the 'particles' present. Dynamic light scattering is sensitive to variations in particle size (in the range of ~1 nm) and interactions of protein molecules in solution (Schmitz, 1990). Dynamic light scattering is routinely used to assess sample mono-dispersity using dilute protein samples (Ferre-D'Amare and Burley, 1994; Bergfors, 1999; Borgstahl, 2007; D'Arcy, 1994). Dynamic light scattering is also used as a research tool for studying protein nucleation and the early stages of crystal growth from crystallizing solutions containing precipitating agents (e.g. Azuma *et al.*, 1989; Georgalis and Saenger, 1993; Malkin and McPherson, 1993; Mikol *et al.*, 1990; Skouri *et al.*, 1992; Wilson, 2003), showing an increase in hydrodynamic radius as supersaturation proceeds (e.g. Malkin *et al.*, 1993; Scheuler *et al.*, 1999).

Light scattering was applied to study the separation of nucleation and growth by Rosenberger *et al.,* (1993) and successfully limited the number of lysozyme crystals grown. However, the procedure involved scintillation and required 50–100 microlitre of sample.

Another study used DLS to monitor the crystallization of proteins mixed with their crystallizing agents in microbatch, so as to get an indication as to when to dilute the trial in order to lead it out of the nucleation phase and into the growth phase (Saridakis *et al.*, 2002). Changes in the aggregation profile of a supersaturated protein sample at nucleation conditions as a function of time were used as an indicator of the induction time for nucleation. DLS showed a significant change in the size-distribution profile of species in solution, which corresponded to the time at which the solution was effectively diluted to metastable conditions, for optimal growth (Figure 5.3).

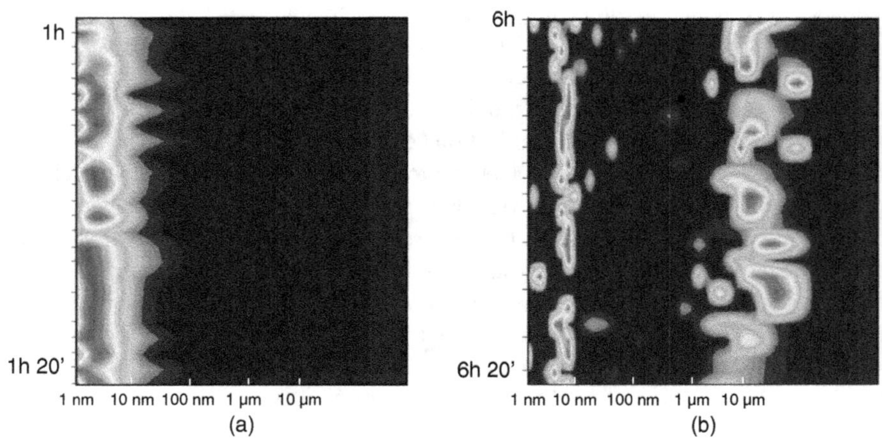

Fig. 5.3. Dynamic light-scattering measurements during a crystallization process. (a) Measurements at 1 h to 1 h 20 min after set up of the experiment showing particles of a size between 1–10 nanometre (b) Measurements at 6 h to 6 h 20 min after set up of the experiment showing particle sizes between 1–10 micrometre indicating that nucleation has taken place. Reprinted from Saridakis *et al.*, (2002) with permission of the authors and the IUCr.

Practical methods such as the separation of nucleation and growth of the crystallization phases identified using light-scattering methods, complement information provided by phase diagrams and can effectively guide optimization experiments to obtain high-quality crystals, especially in those cases where standard methods fail.

5.3 Chapter Summary

This chapter has highlighted the methods for the manipulation of crystallization conditions based on the protein's solubility phase diagram. Several experimental methods for separating nucleation and growth were described. Many of these techniques are simple to execute and require very little sample. Other techniques based upon both static and dynamic light scattering are quantitative and sophisticated, but they require dedicated apparatus. The coming years promise to bring advances in optimization techniques that will play a major role in raising the success rate of producing high-quality crystals and subsequent structures.

6

Strategies to apply when high-quality crystals cannot be obtained

6.1 Introduction

When exhaustive efforts fail to produce crystals from a particular sample, there remains only one logical approach, i.e. to change the sample in some manner. Newman (2006) reports on the underutilization of chemical modification techniques with respect to protein crystallization. Qiu and Janson (2009) build on this adding further techniques and summarize these as: the study of orthologues or homologues; the use of interacting ligands that can be added to the crystallization mix to stabilize the sample; exploring crystallization of the native protein–protein or protein–nucleic acid complex; the use of antibody fragments; and chemical and biochemical modifications. Perhaps the most important method they describe is that of molecular biology. Making protein modifications through site-directed mutagenesis is a powerful method that allows regions of the protein to be removed, or altered deleting flexible regions or altering the surface to promote key crystal contacts. Evdokimov (2007) considers these disparate approaches to protein modification by dividing them into three categories grouped according to the level of time and effort involved. These are (from least to most effort) non-covalent modification of the sample or crystallization conditions, covalent modification of the sample, and the use of mutagenesis, domain refinement and homologues. We discuss some of these approaches here.

6.2 Non-covalent Modification of the Sample

Screening and optimization through different biochemical or biophysical conditions was discussed earlier in the book. Approaches that target the macromolecule directly include the use of crystallization additives, e.g. inorganic ions, neutral and charged substances, detergents, *etc.* and the use of molecules known to bind to the target of interest, e.g. substrates, cofactors, inhibitors or other ligands. In the first case, small molecules promote macromolecular contacts. The macromolecule selects a small molecule from the reagents and the small molecule often appears in the crystal structure at a location that promotes structural stability (Larson *et al.*, 2007; McPherson and Cudney, 2006). These have been commercialized into screens, called 'silver bullets'™. The 'silver bullets' in these screens consist of 200 small molecules, combined to form reagent mixes containing 3 to 20 different chemicals. The cocktails are formulated with one of two precipitating agents; 30% PEG 3350, or 50% Tacsimate™, both buffered at pH 7.0. Similarly, the second approach, substrates, cofactors, inhibitors or other ligands, has also been commercialized in the form of a 'silver bullets™ bio' crystallization kit. These include small organic molecules, organic salts, acids, biologically active molecules, cofactors and ligands, amino acids, peptides, nucleotides, pharmaceuticals, carbohydrates and biochemical pathway intermediates. The development of this area is well described in a series of recent papers (Larson *et al.*, 2007; McPherson and Cudney, 2006; McPherson *et al.*, 2007).

Non-covalent modifications are the simplest first approach as they utilize the existing sample and require minimal effort for reformulation.

6.3 Covalent Modification of the Sample

Covalent modifications may involve significant additional steps in the sample manipulation and crystallization process. They include the chemical modification of groups in the sample, the incorporation of heavy atoms, or limited proteolysis. We discuss a few examples below.

6.3.1 Reductive methylation

Reductive methylation of proteins has been successfully applied to obtain high-quality crystals (D'Arcy *et al.*, 1999; Kim *et al.*, 2008; Rypniewski *et al.*, 1993; Shaw *et al.*, 2007; Walter *et al.*, 2006). Several studies, (e.g. Means and Feeney, 1995; Walter *et al.*, 2006) have demonstrated that methylating the solvent-exposed ε-amino group of lysines changes protein properties (pI, solubility and hydropathy) (Canaves *et al.*, 2004; Kyte and Doolittle, 1982), which may promote crystallization via improved crystal packing. Reductive methylation of proteins is a simple, generic method; it is fast, specific and requires few steps under relatively mild buffer and chemical conditions and can be executed for several proteins in parallel, making this approach an especially viable alternative for most crystallization laboratories.

Kim *et al.* (2008) investigated the application of reductive methylation on a large scale. They applied a previously described reductive methylation protocol (Rypniewski *et al.*, 1993; Walter *et al.*, 2006) to 370 sequence-diverse proteins selected from protein families that had no structural homologues with >30% sequence identity. They expressed 370 recombinant proteins and purified them using standard methods followed by standard crystal-screening methods. Of the 370 proteins, 269 proteins had not previously yielded crystals suitable for structure determination (crystals were too small, poorly ordered, twinned, highly mosaic or multiple), 85 proteins had previously failed to crystallize and 16 proteins were a reference set (not previously screened for crystallization). After reductive methylation, they obtained diffraction-quality crystals for 40 of the 370 proteins, and so far solved 26 crystal structures (Kim *et al.*, 2008). The native and methylated structures of 4 proteins were compared. A decrease in the isotropic B factor was observed and the resolution of the methylated structures was better than that of their native counterparts.

6.3.2 General chemical modification

Newman (2006) notes that "There are many reagents that will specifically modify particular sidechains (Lundblad, 2005); given the wealth of possibilities, it is perhaps surprising that only the reductive alkylation of lysines appears regularly in the crystallization literature (Rayment, 1997; Schubot and Waugh, 2004). A simple chemical modification to free cysteines is quite likely to occur just by adding the reductants β-mercaptoethanol or dithiothreitol Begg and Speicher (1999)". Chemical modification is a technique that offers promise but has not been fully exploited to date.

6.4 Mutagenesis, Domain Refinement and Homologues

Molecular biology approaches offer a number of salvage strategies including surface entropy mutagenesis, removal of disordered regions, truncations, systematic investigation of domain combinations and orthologue and homologue approaches.

6.4.1 Surface-entropy reduction

Crystallization is very sensitive to entropy changes involving both the solvent and the macromolecule. Crystal contacts that order surface residue sidechains result in a loss of entropy (Derewenda and Vekilov, 2006). Rational surface mutation aims to replace exposed polar amino acids, e.g. lysines and glutamates, with amino acids having smaller sidechains, e.g. alanine, reducing surface entropy. Initial experiments used Lys to Ala, Glu to Ala and Glu to Asp mutants and confirmed that mutated proteins were more likely to crystallize (Longnecker *et al.*, 2001; Mateja *et al.*, 2002). As an example of one such study Czepas *et al.* (2004) used Lys to Arg mutations. Several mutations of the globular domain of human RhoGDI were generated, including the single mutants K105R, K113R, K127R, K138R and K141R, the double mutants K(98,99)R and K(199,200)R and the triple mutants K(98,99,105)R and K (135,138,141)R. It was shown that Lys to Arg mutants are more likely to crystallize than the wild-type protein, although not as likely as Lys to Ala mutants. Out of the nine mutants tested, five produced diffracting crystals, especially a K(199,200)R double mutant, which crystallized in a new space group and exceeded by ~1.0 Å the resolution of the diffraction of the wild-type crystal, i.e. reaching 1.6 Å resolution in the final refined structure. Major crystal contacts in the new lattice were created by the mutated epitope.

6.4.2 Proteolysis

It has been shown that fragments or domains of proteins can either crystallize better, or form more ordered crystals compared with the intact protein (Koth *et al.*, 2003; Swatkoski *et al.*, 2007; Wernimont and Edwards, 2009). The advent of biological mass spectrometry has enabled the detection of protease fragments of proteins. Recombinant protein expression has enabled the preparation of proteins that do not contain the protease sensitive domain. By cloning the region(s) of the gene that corresponded to the protease-resistant domain(s), the stable fragment(s) can be expressed in recombinant form, purified to homogeneity and then crystallized. The same principle is also used to identify well-behaved fragments of proteins with known domain structure; in this case the strategy is to identify the approximate boundaries of the domain of interest using sequence alignment, to screen many recombinant versions of this domain for expression (differing slightly in their N- and C-terminal boundaries), and then to select the fragment(s) that can be produced in soluble form for purification and crystallization (Dong *et al.*, 2007). Dong and coworkers tested the general applicability of *in-situ* proteolysis to form protein crystals suitable for structure determination. They added a protease (chymotrypsin or trypsin) digestion step to the crystallization trials of 55 bacterial and 14 human proteins that had proven recalcitrant to their best efforts at crystallization or structure determination. Thus far, they have determined structures of 9/55 bacterial proteins and the human aminoimidazole ribonucleotide synthetase (AIRS) domain (Dong *et al.*, 2007). McPherson *et al.* (2004) have developed an *in-situ* proteolysis approach where the addition of trypsin to drops produced crystals of the sample (bovine lutenizing hormone) not seen without the trypsin present.

6.4.3 Orthologues and homologues

Orthologues are similar genetically coded proteins in different species that have a close common ancestor. Homologues also have sequence similarity but have no close common ancestor.

In cases where the target sample cannot be crystallized, homologues and orthologues offer an alternative approach. A nice example is illustrated with the case of Na^+/Cl^--dependent neurotransmitter transporters (Yamashita *et al.*, 2005). Here, structural information on a human system was unavailable but a bacterial homologue was successfully crystallized. Overall sequence identity may be poor between species but there can be clusters of high sequence conservation. Homologue and orthologue approaches have been used successfully for a long-time in crystallography, for example advice from McPherson (1999b) on targeting a rabbit protein if a bovine case failed. Given the explosion of genomic knowledge becoming available, homologue and orthologue approaches are becoming increasingly powerful.

6.5 Antibody Fragments

Co-crystallization with antibody fragments has been reviewed by Kovari *et al.*, (1995) and used successfully in the case of membrane proteins (Hunte and Michel, 2002) and as crystal chaperones for RNA (Ye *et al.*, 2008). It has also been utilized as a means to determine structure (Tereshko *et al.*, 2008) with additional Met residues providing high-quality phasing information through SAD methods. While successful in a number of cases there are several steps in the process that limit the general applicability, i.e. the selection of the antibody, preparation of a homogeneous species and formation of the complex.

6.6 Chapter Summary

Modification of the macromolecule is an effective approach to crystallize recalcitrant samples. The amount of sample and effort required to undertake these modifications often dictates their practicality. These methods vary widely from the addition of small molecules for reformulation, to altering the protein's chemical makeup using chemistry or molecular biology.

7

Membrane proteins

7.1 Introduction

Approximately 25% of all proteins are membrane proteins and yet less than 200 unique structures are available (Carpenter *et al.*, 2008). The original structures were derived by electron microscopy at 7 Å resolution (Henderson and Unwin, 1976; Henderson and Unwin, 1975). In this case bacteriorhodopsin showed seven transmembrane alpha helices supporting the idea that alpha helices but not beta sheets are the secondary structure of choice for membrane proteins. This idea strongly influenced research on the structure–function relationship of membrane proteins. The beta-barrel models for bacterial porin and beta helix structure of Gramicidin are exceptions to this rule. The first X-ray high-resolution structure of a membrane protein was the bacterial reaction centre (Deisenhofer *et al.*, 1984) again showing alpha-helical transmembrane segments. Membrane proteins are widely regarded as difficult to crystallize with several books dedicated to the topic (Grisshammer and Buchanan, 2006; Hunte *et al.*, 2002; Iwata, 2003). The structural studies of membrane proteins lag those of their soluble counterparts. Nevertheless, there is an acceleration of structural knowledge comparable with that seen for soluble systems (White, 2004) and we can expect knowledge in this area to increase rapidly over the next decade. We will present an overview of membrane protein crystallization and refer the reader to these more specialized texts and references therein for detailed coverage of this challenging discipline.

7.2 Crystallization

Why are membrane proteins under-represented structurally? They are inherently difficult to over-express, and solubilize from their natural environment, a lipid bilayer, typically employing detergents to maintain the protein's native conformation and activity. This water-soluble protein detergent complex is the subject of crystallization trials. The detergent forms a belt-like surface around the protein's hydrophobic surface. The polar regions of the protein extend beyond the detergent belt allowing for protein–protein interactions, required for crystallization. Choice of the detergent, and its concentration relative to the protein are important variables for crystallization

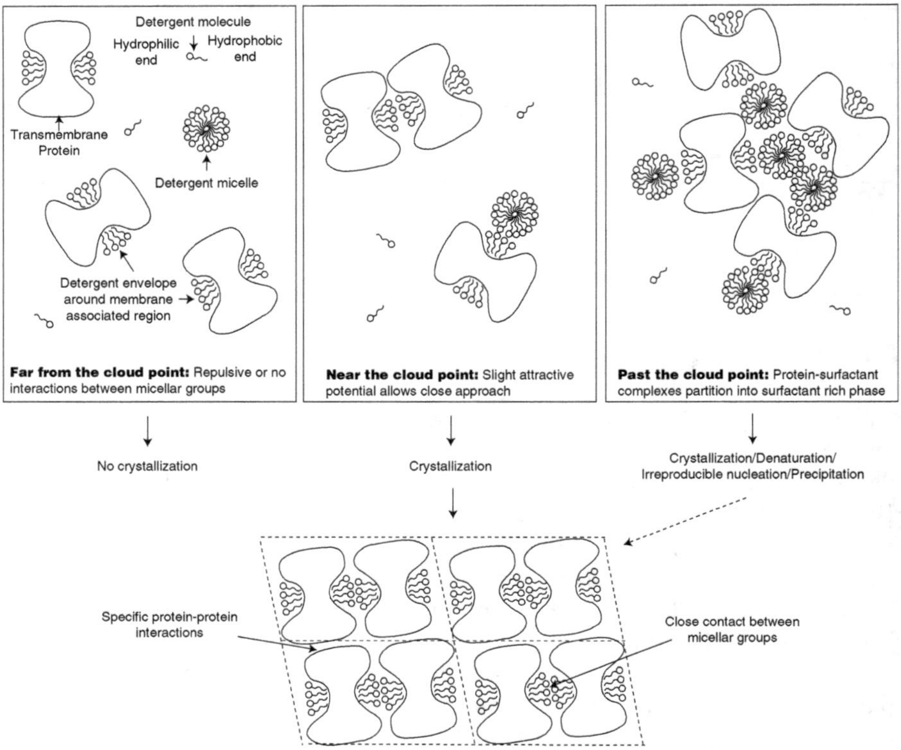

Fig. 7.1. Schematic representation of the effects of micelle–micelle interactions upon the crystallization behaviour of membrane protein–detergent complexes. Far from the cloud point (left panel), forces between the surfaces of micellar groups on adjacent complexes are not attractive, precluding close contacts. Detergent micelles, individual detergent molecules and proteins with detergent shielding the membrane spanning region are present but do not interact. Once the consolute boundary is crossed (right panel), the solution separates into two phases, with the membrane proteins partitioning into the detergent-rich phase. This may denature the protein and/or lead to precipitation; in addition, irreproducible nucleation may result from the presence of two distinct liquid phases. Only when conditions approach but do not exceed the cloud point (centre panel) can the small inter-micellar attractive forces allow close approach of the complexes to one another without the negative effects of phase separation. From Loll *et al.* (2001) with permission of the authors and Elsevier.

of membrane proteins. The *critical micelle concentration* (CMC) is the concentration of detergent above which micelles will spontaneously form. Detergents that are most successful for membrane protein crystallization form small monodisperse micelles, and have low CMC values (millimolar) (Garavito and Ferguson-Miller, 2001).

This is illustrated in Figure 7.1 adapted from Loll et al. (2001). Loll et al. looked at the detergent cloud point (the point at which detergent micelles aggregate and essentially precipitate from solution) in relation to membrane protein crystallization. They made note of observations that many membrane–protein detergent complexes crystallized near this cloud point. The attractive potentials of the micelle–micelle interactions have to be just strong enough to promote packing but not overly strong for that packing to become random and the sample to denature, aggregate or precipitate.

There are two principal types of membrane–protein crystal; type I that form stacked arrays of two-dimensional layers with hydrophobic contacts between the membrane-embedded portions of the protein and hydrophilic contacts between the domains exposed to aqueous solvent, and type II with lattice contacts that involve polar domains of the protein. Most membrane proteins that diffract X-rays fall into the latter category.

7.2.1 Screening with detergents

One approach to crystallizing these proteins has been a conventional screening approach incorporating detergent to solubilize the protein aiming to approach the cloud point but not exceed it, Figure 7.1. Empirically derived detergent phase data for 11 detergents, 11 salts, and 10 PEGS were recently published (Koszelak-Rosenblum *et al.*, 2009). The data were used to generate a 1536 condition crystallization screen that was reported to successfully crystallize 12 out of 14 membrane proteins, 10 of which had never been reported to crystallize. An excellent article is available by Wiener (2004) that offers practical suggestions for successful crystallization based upon detergent phase boundaries. Historical surveys of previous membrane protein crystallization successes have lead to specialized screens (e.g. Horsefield *et al.*, 2003; Iwata, 2003) which have been commercialized, e.g. the MembFac (Hampton Research, USA) Memstart and Memsys (Molecular Dimensions, UK). Similar to soluble proteins, the most commonly used method for crystallizing membrane proteins is vapour diffusion, however, microbatch, microfluidics and lipidic cubic phase or lipidic mesophase have also been used successfully. An increasing number of membrane proteins using a variety of different detergents have been crystallized using the microbatch under oil technique. Some of these had failed to crystallize by other methods (e.g. Henkamer *et al.*, 1992; Snijder, 2003; Stocker *et al.*, 2005; White, 2004).

7.2.2 Lipidic cubic-phase crystallization

The most significant factor, which is unique to the crystallization of membrane proteins, is the necessity to remove them from their native lipid bilayers by detergent solubilization. This exposes the proteins to an environment with considerably different properties from their native membranes. In order to overcome this major problem and keep membrane proteins in a quasi-solid membrane environment during the crystallization process, Landau and Rosenbusch (1996) introduced the use of bicontinuous cubic phases of lipids for membrane–protein crystallization. This membrane system, consisting of lipid, water, and protein in appropriate proportions, forms a structured, transparent, and complex three-dimensional lipidic array, which is pervaded by an intercommunicating aqueous channel system. Such matrices provide nucleation sites that the authors call 'seeding' and support growth by lateral diffusion of protein molecules in the membrane. They successfully obtained crystals of bacteriorhodopsin in the bicontinuous cubic phases. The use of lipidic cubic phases has been extensively reviewed by Caffrey *et al.* (2003).

Several years later the use of cubic phases were successfully miniaturized and automated. Miniaturization was achieved with standard microwell plates utilizing nanolitre (200 nl) quantities of lipidic cubic phase (Cherezov and Caffrey, 2003; Nollert, 2002). Diffracting crystals of bacteriorhodopsin were obtained using this procedure (Nollert, 2002). A robotic system for crystallizing membrane proteins in lipidic mesophases was reported (Cherezov *et al.*, 2004) reducing experiment volumes to the picolitre scale. A significant breakthrough was lipidic cubic-phase crystallization and structure determination of medically important G protein-coupled receptors (Cherezov *et al.*, 2007; Jaakola *et al.*, 2008). More recently a liquid analogue of the cubic phase called sponge phase (Johansson *et al.*, 2009) has been introduced and commercialised.

7.2.3 Antibody fragment approaches

Fab and Fv antibody fragments have been used to extend the available polar surface of membrane proteins to form crystal contacts (Hunte *et al.*, 2000; Ostermeier *et al.*, 1995; Zhou *et al.*, 2001). Antibody fragments suitable for cocrystallization should bind to the protein in its native conformation, have high binding affinity and most importantly bind to a discontinuous surface (Carpenter *et al.*, 2008).

7.3 Characterization Using Neutron-Enhanced Contrast

Separate from crystallization trials per se, neutron crystallography can be used to determine how the detergent packs within membrane–protein crystals. Such studies (reviewed by Timmins (2006)) demonstrate that as well as protein–protein interactions being important in the formation of crystals they also show protein–detergent and even detergent–detergent interactions can be crucial. Using the known X-ray structure and neutron diffraction data measured at a number of H_2O/D_2O component contrasts (Figure 7.2) it is possible to calculate neutron scattering length density maps that show only the detergent. This is illustrated in Figure 7.3 where the detergent network in this case for crystals of monomeric outer membrane phospholipase A is shown (Snijder *et al.*, 2003).

7.4 Chapter Summary

The crystallization of membrane proteins is complicated by difficulty in their overexpression, extraction from a native lipid bilayer, maintaining protein conformation

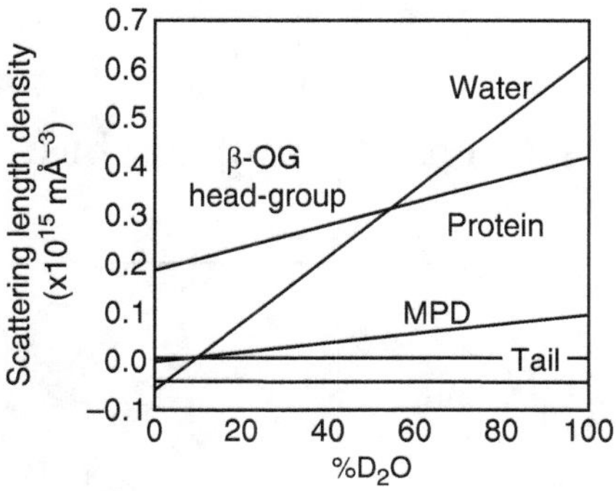

Fig. 7.2. Scattering-length densities as a function of D_2O concentration are plotted for water, a typical protein, tail region of β-OG, head group of β-OG, and 2-methyl-2,4-pentanediol (MPD). From Snijder *et al.* (2003) with permission of the authors and the *J Struct Biol.*

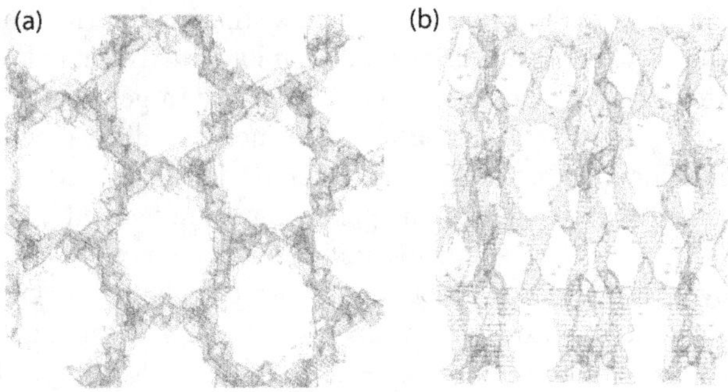

Fig. 7.3. Details and overview of the detergent structure with two perpendicular orientations of the neutron map showing that the detergent forms a continuous network that extends throughout the entire crystal (Snijder *et al.*, 2003) with permission of the authors and the *J Struct Biol.*

and activity after extraction, and the difficult task of crystallizing the protein detergent complex. Although ~30% of the genes in known genomes code for membrane proteins, there are only 193 unique structures to date. A few of the most popular approaches for crystallization of these challenging targets have been presented, along with references offering more detailed descriptions of the methods.

8

Alternative approaches

8.1 Gel Growth

A well-known method for optimizing the size and quality of crystals is by growing them in gelled media (Garcia-Ruiz, 1994; Garcia-Ruiz, 2003a; Garcia-Ruiz *et al.*, 2001; Gavira and Garcia-Ruiz, 2002; Gavira *et al.*, 2002). The presence of gel in the crystallization drop can reduce convection and sedimentation (in some ways mimicking crystallization experiments carried out in microgravity but at lower cost in the controlled environment of a home lab). Crystallization in gels can be performed in batch, microdialysis, vapour diffusion and free interface diffusion (FID) (Bernard *et al.*, 1994; Biertumpfel *et al.*, 2002; Cudney *et al.*, 1994a; Robert and Lefaucheux, 1988; Sica *et al.*, 1994) as well as in the gel acupuncture method where the crystals grow inside an X-ray capillary that is inserted into a gel directly (Garcia-Ruiz, 1994) or via an oil layer (Moreno *et al.*, 2002). The use of gels for the crystallization of proteins was reported by Low and Richards (1954) and the method has been under-utilized. This may be due to the relatively complicated procedures required when applying gels to crystallization trials and also to the large quantities (mostly $\geq$ 10 μl) of sample needed.

However, the last several years have seen major improvements in the use of gels by way of miniaturization and automation. As an example, the Granada Crystallization Box (Garcia-Ruiz *et al.*, 2002) enables the growth of crystals in 1-μl volumes of gel inside long narrow capillaries such that "counter-diffusion" crystallization is possible. An even simpler way is to set up gelled trials in microbatch, which for the first time, has enabled the automatic dispensing of submicrolitre gelled drops in a high throughput mode (Chayen and Saridakis, 2002). The experiments are performed using a computer-controlled dispensing system in which precipitant, buffer, protein and additives are put into different syringes. By placing a gel solution while it is still in liquid form, into one of the syringes, it is possible to automatically dispense microbatch trials that form the gel/crystallization mixtures in final volumes of 0.3–2.0 μl. This can be achieved during a comparable timescale and with the same ease as conventional automated microbatch trials. Comparison of crystals grown in gels with those grown in standard trials show an improvement in crystal order as measured from X-ray diffraction properties (Dong *et al.*, 1999). Harvesting crystals from the gelled drops is performed in the same way as from standard microbatch trials (Chayen, 2005).

8.2 Microgravity

In microgravity, an ideal, diffusion-only, condition can be established during crystallization, eliminating density-driven turbulence that is typical at $1g$. Microgravity has been used as a variable for a wide variety of crystallization experiments to evaluate the resulting protein crystal quality, for fundamental studies and, in more specialized applications, for producing homogeneous crystal-size distributions. Overall, microgravity has delivered bigger and more perfectly formed macromolecular crystals as reviewed in detail by Snell and Helliwell (2005). However, the use of microgravity for macromolecular crystal growth has been a subject of intense debate. Disadvantages of microgravity include a lack of access, high costs per crystallization experiment (albeit mitigated by the small weight of the apparatus required), strict safety regulations and established protocols that are not easily altered, and difficulty realizing true microgravity conditions – astronaut activity for example causes periods of gravity-like effects 'g-jitter' (which can be overcome on unmanned systems). Moreover, the inevitable high profile nature of the experiments places considerable strain on the research and development of how to exploit the space environment optimally. Although space has provided 'benchmark' results, opportunities that are solely earth based, such as gel growth described above, microfluidics, and electromagnetic effects described below, are relatively attractive, if not at least more readily attained for the majority of investigators.

8.2.1 Interferometry

Microgravity has stimulated a strong activity in monitoring of crystal growth through CCD video and also interferometric methods monitoring refractive-index changes of the solution. As techniques, simple video imaging and interferometry have also confirmed the guiding principle of optimization of crystal growth under quiescent conditions by promoting the formation of a depletion zone around the growing crystal. Microgravity is not 'zero-gravity' and 'g-jitter' disturbances from, e.g. equipment vibration and/or astronaut movements during, exercise periods can show up as a disturbance of the depletion zone. This subsection summarizes these methods and some of the key results.

Michelson interferometry has been applied to determine the solubility of a protein within the first several hours (Kurihara *et al.*, 1996). Seed crystals of a chosen protein were grown in a cell designed for interferometric observation (Komatsu *et al.*, 1993; Miyashita *et al.*, 1994). The fringes observed by the interferometer appear around the crystals (bent upwards or downwards with changes in temperature), corresponding to either growth or dissolution of the crystals and remained flat at the equilibrium temperature for specific chemical conditions. Using such interferometric techniques one can judge whether a solution is under- or super-saturated at each condition.

Several types of interferometer schemes are applicable to monitoring crystallization (Shlichta, 1986). Two successful types of interferometry that have been applied to monitoring crystal growth are Mach–Zehnder and Michelson interferometry. The

Mach–Zehnder interferometer is suited to monitoring the global solution changes within the crystallization chamber and changes in concentration of solvent/protein around the growing crystal. The Michelson interferometer is used for direct measurement of growth of the crystal surface. The sensitivity of both types of interferometry can be greatly increased by phase shifting (Burning *et al.*, 1974; Komatsu *et al.*, 1993) and with suitable design of the crystallization reactor, can both be used simultaneously.

In the Mach–Zehnder interferometer any change in concentration of the solution will affect the refractive index producing interference fringes between the measuring and reference beams. The continual monitoring of such fringe patterns has been used to observe the crystallization of lysozyme (Boggon *et al.*, 1998; Snell *et al.*, 1996). By monitoring the rate of interference fringes passing through the crystallization chamber the concentration of salt as precipitant and the protein in solution can be determined over time. With the addition of phase shifting it is possible to measure the concentration boundary around the growing crystal (Onuma *et al.*, 1993) although for this kind of observation the thickness of the solution layer should not be too great (Komatsu *et al.*, 1993).

The Michelson interferometer is constructed such that the measuring beam is reflected by the growing crystal surface. Any change in the crystal surface will alter the distance the light travels, causing shifts in interference fringes with respect to a reference beam. This is a very sensitive technique for measuring vertical displacement. Chernov *et al.* (1998; 2002) pioneered the use of Michelson interferometry for studying the kinetics of crystal growth It has also been used very successfully in studies of lysozyme (Vekilov *et al.*, 1995) concanavalin A, thaumatin and turnip yellow mosaic virus (Kuznetsov *et al.*, 1995).

8.2.2 Depletion zone

Video and interferometric techniques have been critical in developing and proving the standard model for understanding the effects of microgravity on macromolecular crystal growth. This is based on the concept of a depletion zone (McPherson *et al.*, 1991). In the absence of acceleration, a crystal is subject to Brownian motion as on the ground, but unlike the ground case, there is no acceleration inducing it to sediment. A growing crystal in zero gravity will not move with respect to the surrounding fluid. Moreover, as macromolecules leave solution and add to the crystal, a region of solution depleted in protein is formed. Usually this solution has a lower density than the bulk solution and will rise upward in a 1*g* field as seen in both small molecule (Chen *et al.*, 1979) and macromolecular crystallization (Figure 8.1) (Pusey *et al.*, 1988). In zero gravity, the buoyancy force is eliminated and no buoyancy-driven convection occurs.

Because the position of the crystal and its depletion zone are stable in microgravity, the crystal can grow under conditions where its growing surface is in contact with a solution that is slightly supersaturated. In contrast, the sedimentation and convection that occur under 1*g* place the growing crystal surface in contact with bulk solution

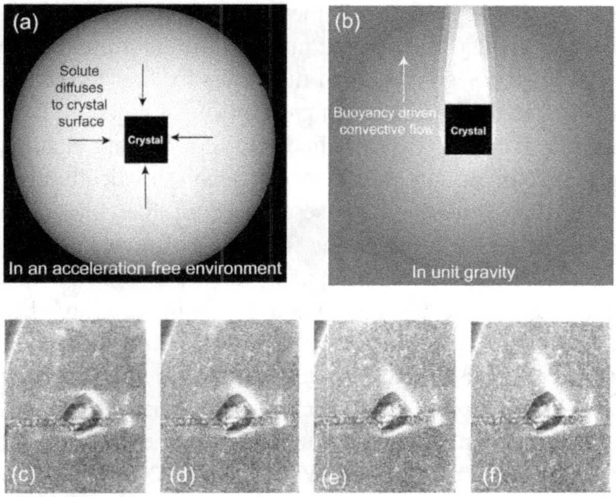

Fig. 8.1. Illustration of the zone of depleted macromolecule around a growing crystal in (a) schematic of an acceleration-free environment compared to the convective plume formed by growth in (b) unit gravity. Also shown (c)–(f) are Schlieren photography images of the convective plume that forms over time from a lysozyme crystal (approximately 1.2 mm in size) grown on the ground (Pusey *et al.*, 1988), the time interval between each image is 12 s giving a plume velocity of approximately 30 $\mu m\ s^{-1}$. Figure taken from Snell and Helliwell (2005) with permission of the IOP.

that is typically several times supersaturated. Lower supersaturation at the growing crystal surface allows more high-energy mis-incorporated growth units to disassociate from the crystal before becoming ideally oriented and trapped in the crystal by the addition of other growth units. However, since microgravity is not in fact zero gravity, the buoyancy-driven convection and sedimentation are only attenuated rather than eliminated. Promotion of a stable depletion zone in microgravity is postulated to provide a better-ordered crystal lattice and benefit the crystal-growth process.

Otalora *et al.* (2001) also observed concentration-depletion zones around growing macromolecular crystals (ferritin and lysozyme) under reduced-gravity conditions. These depletion zones, ideally having radial symmetry, were distorted by crystal and liquid movements during microgravity macromolecular crystal-growth experiments as a consequence of buoyancy-driven convection triggered by residual accelerations and g-jitters, i.e. achieving true zero gravity is very difficult.

8.3 Microfluidics

The application of microfluidics to crystallization allows for hundreds to thousands of fluid-routing and mixing events to be conducted combinatorially within a single integrated device using only tiny amounts of sample.

Microfluidics can be applied to all methods of crystallization. Ismagilov and colleagues (Zheng *et al.*, 2003) described microfluidics to mimic microbatch crystallization under oil. Their design contained fluid lines for four different solutions. The fluid lines were arrayed to meet and combine at a single junction. By controlling the amount of fluid that is pulsed by each line into the merging chamber, the various ratios of sample, precipitant, and buffer could be systematically altered. The composite solution is then routed into a separate fluid line that contains a water-immiscible fluorinated oil to promote the formation of reverse micelles and permit solute mixing by chaotic advection (Song and Ismagilov, 2003; Thorsen *et al.*, 2001). Once set up, the droplets are allowed to equilibrate in the chip where they can be visualized to assess crystal growth (Berger, 2007).

The Ismagilov group also described a variation on this device for setting up 'vapour diffusion-like' experiments. The chip uses a small plug of water-permeable oil to separate a drop containing protein and precipitant from another solution containing a high concentration of salt (Zheng *et al.*, 2004). Water migrates from the target/precipitant drop into the high-salt solution, thereby dehydrating the drop and increasing the solute concentrations to promote crystal nucleation and growth.

However the big breakthrough is in the miniaturization and automation of the free interface diffusion method that has been made with the development of microfluidic chips that can now be used for very high throughput (Hansen *et al.*, 2002). Up to eight proteins under 96 conditions can be screened (with the built-in fine tuning of supersaturation provided by FID) simultaneously in the TOPAZ 8.96 chips (Fluidigm Corp., San Francisco, USA). The protein and precipitant solutions are loaded into the chips and put into contact via microfabricated valves (Hansen *et al.*, 2002; Lau *et al.*, 2007). The power of the method lies not only in the miniaturization, which leads to minimal protein consumption and maximal speed, but also in the special physics of fluid flow and mass transport in microchannels (Chayen and Saridakis, 2008), such as the dominance of laminar over turbulent flow, large surface/volume ratio, *etc.* (Hansen and Quake, 2003). Experiments using microvalve-controlled microfluidic chips, have been reported that have resulted in increased numbers of 'hits' compared to conventional screens (Lau *et al.*, 2007; Sommer and Larsen, 2005). There is no doubt that microfluidics is very successful but it currently requires dedicated and relatively expensive hardware and consumables.

8.4 Magnetic and Electric Fields

High magnetic field facilities have been steadily expanding along with the provision for materials processing applications. High field strengths, tunable settings and choice of inhomogeneous and homogeneous fields have brought this technology to a stage where it can be applied to macromolecular crystallization. Heijna *et al.* (2007) have recently demonstrated a versatile and accessible way to realize an earth-based tunable-gravity environment for crystal growth using the High Magnetic Field Facility in Nijmegen in the Netherlands (www.hfml.ru.nl). Within an inhomogeneous magnetic field protein crystal-growth conditions such as convective flow can be controlled while simultaneously visualizing the growing crystal and its surrounding fluid.

As described in Section 8.2 microgravity has been the method used to remove buoyancy driven fluid flows in crystal growth solutions. Crystal growers have harnessed microgravity in an attempt to create a quiescent, i.e. motionless and dormant fluid state to promote both the nucleation and the crystal-growth stages, unaffected by density-driven fluid flows. It also removes the sedimentation effect of crystals more dense than the mother liquor, which occurs under gravity. High magnetic and electrical fields can be used to mimic the effects seen in microgravity (Al-Haq *et al.*, 2007a; Astier *et al.*, 1998; Banci *et al.*, 2000; Lin *et al.*, 2000; Saijo *et al.*, 2005b; Tanaka and Ataka, 2002; Wakayama *et al.*, 2006).

High magnetic fields can be harnessed in two ways according to whether a homogeneous field (HF) or an inhomogeneous field (IHF) is used. IHF, the method of relevance to Heijna *et al.,* (2007), involved a magnetic field gradient due to its divergence.

Such fields can be applied even to a 'diamagnetic' material, which is a material with no unpaired electrons, unlike a paramagnetic material, whereby the electrons in its atoms are sufficiently displaced with respect to the positive nuclei to create an induced magnetic dipole moment. A vertically diverging magnetic field causes a resultant, i.e. non-zero force in the vertical (but a zero-resultant force in the horizontal). By tuning the magnetic field, B_z, and its gradient, B_z', and setting the resultant vertical force to balance the gravity force then a zero-gravity fluid condition can be established. The precise tuning value depends specifically on the nature of the fluid and its concentration. The control of the fluid in this way is different from the condition for levitation of the crystal and the avoidance of sedimentation, because of the different magnetic susceptibilities of the crystal and the fluid. Hence, whilst in microgravity, convection and sedimentation are reduced simultaneously, this is not the case for an IHF tunable-gravity experiment. This can be viewed in two ways. First, it is a limitation of the magnetic field versus microgravity. However, secondly, the IHF allows the conditions of 'convection free' versus 'sedimentation free' to be separated out and their relative importance in protein crystal growth evaluated in detail.

There are some limits of the IHF approach in evidence at present. An earlier paper from Vlieg and his collaborators (Poodt *et al.*, 2006) studied inorganic crystal growth and explored the accuracy of the settings within the growth chamber. Even at the IHF condition balancing gravity there were some residual fluid flows equivalent to $\sim$0.5 micrometre/s. This is similar to the limits in microgravity where vibrations on the space platform (e.g. the Space Shuttle or a Space Station) create g-jitter, which can cause small movements of the fluid disturbing the quiescent state desired (Boggon *et al.*, 1998).

It is important to also mention the applications of magnetic homogeneous fields (HF) and of electric fields (EFs) in protein crystallization. Saijo *et al.,* (2005a) used a 10T HF to show improved protein crystal growth and improved order of charged residues on loops on their test protein lysozyme. The magnetic 'Lorenz' force on moving electric charges served there to restrain the charged amino acid residues during crystal growth, and that were stable after removal of the HF, i.e. as the crystal had to be removed for X-ray data collection at the synchrotron. Sazaki and Ataka have been pioneers of the use of magnetic fields for protein crystal growth (Ataka *et al.*, 1997; Sazaki *et al.*, 1997). Most recently, the use of combined magnetic and electric fields has achieved a quite remarkable homogeneity of protein crystal size and shape. The use of EFs in protein crystal growth has recently been reviewed by Al-Haq *et al.,* (Al-Haq *et al.*, 2007b).

8.5 Chapter Summary

The control of fluid flows in protein crystal growth has had benefits, as we have documented in this chapter. The control of the fluid physics state using gels, microgravity and microfluidics as well as with applied magnetic and electric fields reveals a strong interest from physicists in the cross-disciplinary effort to turn macromolecular crystal growth firmly into a science.

PART III

DIFFRACTION

9

Experimental aspects

9.1 The Diffraction Pattern

The diffraction pattern obtained reveals many details about how problematic the following data-processing step may become. To obtain the best possible structural information you need the best diffraction data. Experimentally there is a wide range of choices to be made in collecting the diffraction data. A good guide to experimental parameters can be obtained from dedicated programs such as BEST (Bourenkov and Popov, 2006), and other software, e.g. MOSLFM (Leslie, 1992), HKL2000 (Otwinowski and Minor, 1997) or D*Trek (Pflugrath, 1999) which have strategy options to maximize the completeness or anomalous signal by appropriate choice of orientation and oscillation parameters.

A careful inspection of the diffraction pattern can reveal features that may pose problems later. Obvious problems include the presence of large ice rings that may mask individual reflections or anisotropic scattering leading to regions of good and poor data. More subtle effects are shown in Figure 9.1. In Figure 9.1(a), the crystal diffracts to high resolution. This can cause dynamic range challenges for the detector used for collecting the data and typically multiple data sets are collected at different detector distances with a reduced flux and/or exposure time for the low-resolution data pass. For the high-resolution data, a large beamstop can be used to attenuate the low-resolution data and thereby avoid overloading of pixels. This overloading can be a significant problem on image-plate detectors and can also cause 'bleeding' of reflections on CCD-based detectors. It is important to note that when multiple passes are required to collect the full dynamic range the low-resolution data should always be collected first. Radiation damage affects the entire crystal and therefore all the data at once – one doesn't perceive it visually in the low-resolution data initially because the reflection peak heights are relatively large. Unfortunately if one collects the high-resolution data first the difference seen in the low-resolution diffraction data, when one comes to collect it, is significant. A low-resolution data set can be collected with the same dose it takes to collect two or three high-resolution diffraction images and can significantly improve scaling results.

Beyond the shadow from a large beamstop there are two features of interest in the diffraction pattern shown in Figure 9.1. Significant diffuse scatter is present that can reduce signal-to-noise and spot elongation can be seen in the highest-resolution

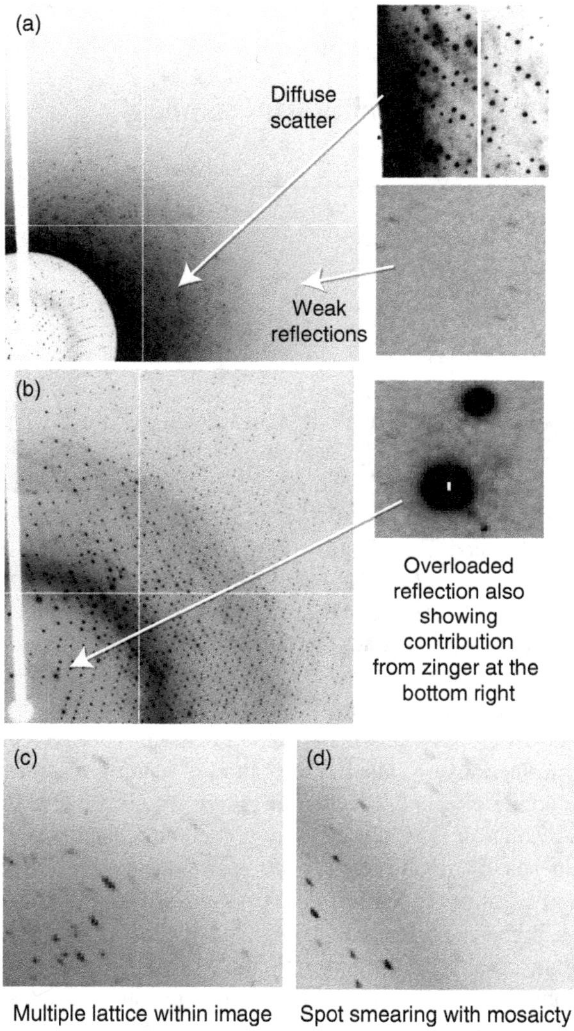

Fig. 9.1. Examples of features seen in diffraction patterns. (a) Ultrahigh-resolution, 0.85 Å data with strong diffuse scatter (both as a general ring and features within the pattern) and weak reflections (note large beamstop and reflection penetration) (b) Low-resolution swathe from same crystal with enlargement of an overloaded reflection (with a peak being cutoff) and a zinger visible to the side of the reflection (c) Multiple lattice (possibly due to multiple crystals in the beam) and (d) High mosaicity.

reflections. We cannot overcome diffuse scatter and, although it provides information, to date we cannot usually make good use of it. Reflection spot elongation can be significant at the extremes of the detector, especially if the detector is translated off-axis to collect high-resolution data. If the unit cell allows we can compensate for this by reducing the wavelength rather than shortening the distance. This reduces the

Bragg angle and therefore geometric smearing. In Figure 9.1(b) the crystal is diffracting very well but closer inspection of the reflections indicates overloads in the low-resolution data. This will be flagged by the data-processing software but ideally should be avoided by analysis of the first image and careful choice of exposure time and oscillation or appropriate beam attenuation. Below and to the right of the overloaded reflection is a zinger. These are caused by the interaction of cosmic radiation with the fibre optic taper in CCD detectors and produce false reflections. They are characteristically formed from a few pixels compared to reflections or may have tail-like shapes, again distinct from reflections. Usually data-integration procedures can avoid them unless they directly overlap with a reflection. Some detectors have the option to 'dezinger' where two identical exposures are taken. The zinger is caused by a transient event and any differences between these two images are thereby removed by comparison.

In Figure 9.1(c), multiple lattices are present within the diffraction image due to a satellite crystal (or crystals) on the main target. If the diffraction from the satellite is not too strong and the lattices are well separated, the data can often still be used. In Figure 9.1(d) the reflections are from a single crystal but smeared (similar to those in (c)). This is an effect of a larger than ideal mosaicity and will be observed in subsequent images where the same reflections will persist.

High mosaicity especially coupled with large unit cells and strongly focusing, wide-bandpass optics often lead to overlapping reflections. With current data-processing software this data cannot be effectively used. Similarly, most current data-processing software makes use of empirical profile-fitting methods. These have no ability to distinguish between coherent and diffuse scattering and model profiles generated in regions of high-frequency background variations or parasitic scattering are poorly determined. At high resolution, large detector incident angles cause significant profile variations over a small resolution range. A bias toward apparently smaller-sized apparent reflections at higher resolution due to systematically lower intensity leads to overestimated B factors and incorrect scaling of low-resolution with high-resolution data. Broadening (smearing) of model reflection profiles occurs when pattern reflections are sampled over too large an area.

We can consider various classes of crystal, physical parameters and detector effects and how they impact the data collected. If we have overlapping reflections, one has to move the detector back to increase the spot separation. Often, when this is the case one loses data due to the finite size of the detector. To overcome this, the detector can be moved back then translated perpendicular to the oscillation axis. This still reduces the overall solid angle of data collected and so the number of images recorded and the reciprocal space sampled will have to increase to take this into account.

If the crystal is split apparently single reflections at low resolution consist of two or more components. At high resolution the components are resolved as separate reflections, which do not scale together. Intermediate-resolution reflections that partially overlap are discarded. For crystals subject to non-merohedral twinning (see Chapter 12) significant partial overlap occurs over large volumes of reciprocal space, degrading model profiles. Where many crystallites are present, multiple patterns result, one from each crystallite. With nearly inevitable overlap it is difficult to determine which crystallite a reflection belongs to and hence produce a suitable profile. Such effects can cause considerable problems in autoindexing. Treating composite reflections as single is often not effective.

The cryo-preservation protocol during data collection (see Chapter 14) introduces other aspects to the resulting X-ray data. The mosaicity is increased (a detrimental effect) but radiation damage effects and atomic motion are reduced, both beneficial effects. Achieving a workable balance between these parameters can be a matter of trial and error and experience.

9.2 Structural Detail

The ultimate product of diffraction is the structure at some level of atomic detail, termed the resolution. For historical reasons this is measured in Angstrom (Å) where 1 Å = 0.1 nm. In the case of X-ray crystallography the information obtained for a particular resolution is described in Table 9.1.

Figure 9.2 demonstrates an example of this in the structure of apocrustacyanin A1 where the data has been truncated at successive resolutions, 1.4 Å, 2.3 Å and 3.2 Å, respectively (Minichino *et al.*, 2003). In X-ray crystallography the structural data is represented by maps of electron density. These are produced by comparing the

Table 9.1. Typical features visible in X-ray crystallographic structures as a function of the diffraction resolution.

Resolution	Details seen
3.2 Å	It becomes possible to trace the polypeptide chain. At resolutions less than this it can be difficult to trace with certainty, e.g. the hydrogen bonding between adjacent strands in beta sheets can lead to a false direction of the chain trace being taken.
2.8 Å	Bound waters become visible. At resolutions worse than this the determination of bound waters is much less reliable.
2.0 Å	At this resolution model refinement results in a precision typically of 0.3 Å rms on atomic positions as indicated by the diffraction precision index (Blow, 2002b; Cruickshank, 1999).
1.5 Å	Split occupancy sidechains start to become well resolved depending on relative occupancy.
1.2 Å	Atomic resolution in the electron density is achieved.
0.9 Å	The X-ray data to parameter ratio is sufficient for chemical crystallography style protein model refinement including anisotropic refinement of APDs and bond-length analysis of the well-ordered carboxy/carboxylate sidechains to decide the protonation state. For histidines such bond-distance analysis remains out of range and protonation determination (as deuterium) requires neutron protein crystallography analysis. Note that the loops of the protein and other more mobile components still must remain restrained to dictionary values and bond length analysis then is not possible (e.g. see Ahmed *et al.*, (2007)).
0.6 Å	The amino acids with $B_{\text{factors}} < 3$ can have their electron densities treated as aspherical and thus valence electron densities can be resolved;

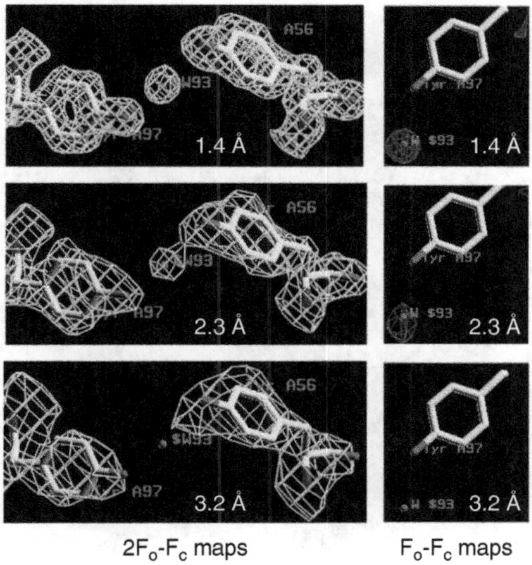

Fig. 9.2. Simple truncation of data from crystals of apocrustacyanin A1 at different resolution ranges. The 2Fo-Fc maps are contoured at 1σ, whereas the Fo-Fc maps are contoured at 2σ with no waters in the model (Minichino *et al.*, 2003) used with the permission of the authors and the IUCr.

experimentally recorded data to that calculated from the current model. In the example given two types of electron-density maps are shown, a 2Fo-Fc map and a Fo-Fc map. These are used to indicate the convergence of the structural model to the experimental data. The electron density is shown by a grid contoured at defined sigma levels and the model as a stick figure surrounded by this grid. A water bridging two tyrosine molecules is clearly visible at the highest resolution. It is still visible at 2.3 Å albeit with electron density starting to smear. By the time the resolution is truncated to 3.2 Å there is no evidence for the water molecule. In the same sequence of images the characteristic ring of the tyrosine starts to blur until it becomes less distinguishable from the main chain. Obviously this is a hypothetical example but still serves to illustrate the effects of truncated resolution.

For pharmaceutical design 2.0 to 2.5 Å is a reasonable resolution where ligands binding to the macromolecule (i.e. potential pharmaceutical precursors) can be resolved in detail. For the understanding of certain mechanisms, especially those involving weakly scattering components like water, then higher resolution is needed. In some cases, X-ray diffraction can reveal hydrogen positions, Figure 9.3, but often these have to be inferred or other techniques such as neutron diffraction used.

In neutron protein crystallography the typical diffraction resolution range of deposited analyses in the PDB are between 3 to 1.5 Å resolution. Since a deuterium atom scatters neutrons as strongly as a carbon atom it is possible to determine the protonation state of, e.g. a histidine at a resolution as 'poor' as 3.0 Å. At 1.5 Å resolution

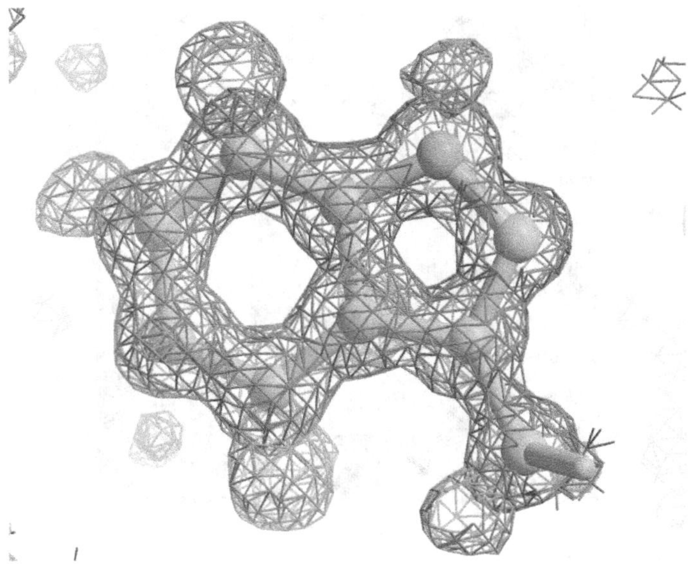

Fig. 9.3. High-resolution X-ray structure of a tryptophan residue in xylose isomerase (Snell *et al.* to be published). The hydrogen atoms were unmodelled and show up in the difference map density (a lighter shade of grey).

the cases of negatively scattering hydrogens, in non-perdeuterated protein, e.g. on phenylalanine rings is also resolvable.

9.3 Determining Structure when the Resolution is Not Ideal

A critical diffraction resolution to achieve in phasing of the X-ray diffraction reflections is at least 3.2 Å resolution. What if the crystal you have only diffracts to say 4 Å resolution but has given reasonable phasing, e.g. either via MAD or MIRAS thus far? You then may find a 3.0 Å resolution crystal but you have the frustration of that crystal not yielding very good phasing, e.g. through lack of isomorphism in the MIR. Repeated efforts to improve your crystallization to yield a higher resolution fail as do your repeated efforts to find a better derivative. Then what?

One possibility is the technique of 'cross-crystal averaging'. By transforming the electron-density map from one to the other crystal form, map averaging has been found to improve the observed map to be adequate to allow model fitting. These procedures are *summarized* for example in Taylor (2003) and used for example in structural studies of Newcastle disease virus haemagglutinin-neuraminidase (HN), a structure plagued with non-isomorphism problems (Crennell *et al.*, 2000). The technique is illustrated in Figure 9.4

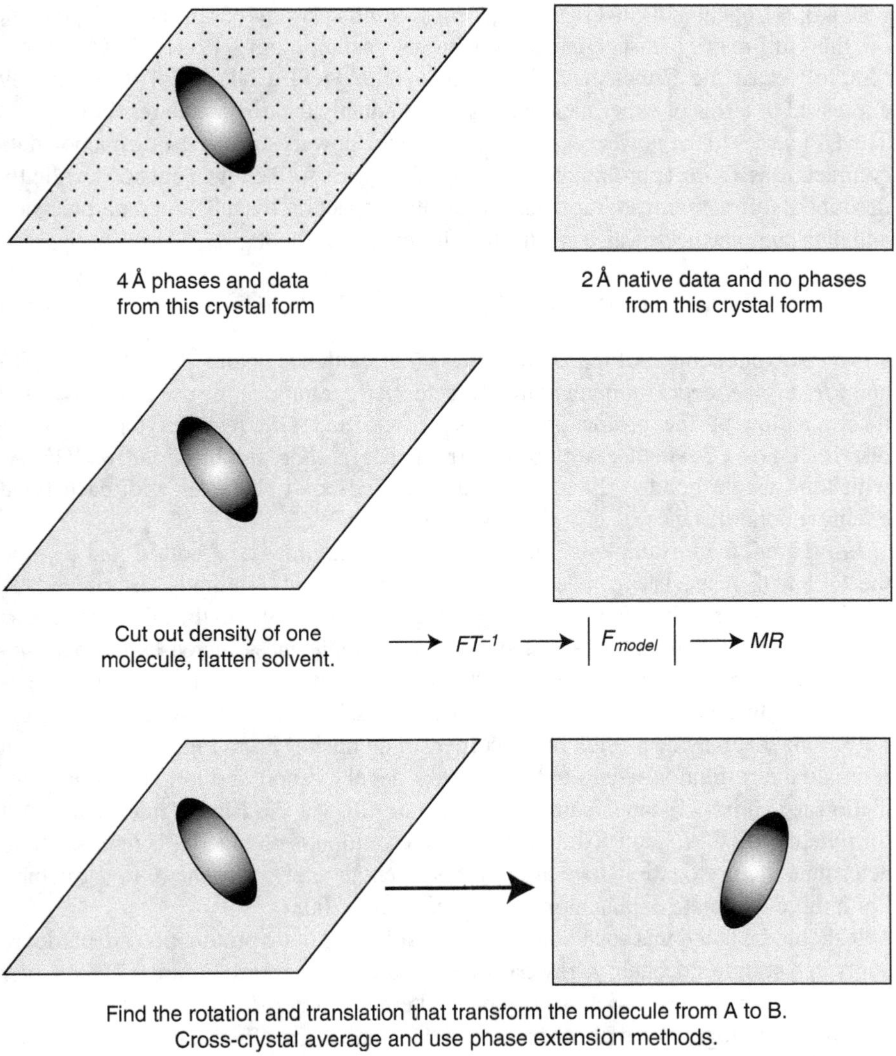

Fig. 9.4. Cross-crystal averaging from Taylor (2003) used with the permission of the author and the IUCr. Two crystal forms of the same macromolecule for which phase information to low resolution is known for one form (on the left) and high-resolution data but no phase information is known for the other (on the right).

9.4 How Accurate is the Structure?

Cruickshank (1999) introduced a way of describing the overall diffraction precision of a macromolecular structure via an index called the diffraction-component precision index 'DPI' which can be applied at any diffraction resolution. The DPI is a quantity notated as $\sigma(x, B_{avg})$, the standard deviation of a coordinate x for all atoms with a B_{factor} equal to the average B_{factor} 'B_{avge}' for the structure. This is not dependent on the 'full-matrix least squares' refinement methodology (Rollet, 1965). Blow (2002b) recast the Cruickshank 1999 analysis to yield a DPI, eqn (9.1), directly expressed in terms of experimental parameters notably the diffraction resolution, d_{min} The DPI varies to the power 5/2 on d_{min}, and to the power −5/6 on the diffraction data completeness, C, i.e. approximately linearly. Thus, as is well known, but now explicitly described, diffraction resolution has a dramatic impact on overall structural precision and data completeness is also worth striving for. The DPI, $\sigma(x, B_{\text{avg}})$, is given by:

$$\sigma(x, B_{\text{avg}}) = 0.18(1+s)^{\frac{1}{2}} V_{\text{M}}^{-\frac{1}{2}} C^{-\frac{5}{6}} R_{\text{free}} d_{\text{min}}^{\frac{5}{2}} \tag{9.1}$$

S is the solvent content of the crystal, V_{M} is the Matthews volume (Matthews, 1968) and *Rfree* is the model refinement free R factor. An example of the use of DPI is in the determination of the protonation states of Asp and Glu residues based on the discrimination of whether a carboxyl group has a short double bond (~1.17 Å) with long single bond (~1.31 Å) versus a mixed-bond character with each bond having a bond distance of 1.25 Å), Figure 9.5.

In Figure 9.6 an example is illustrated showing the impact of both C and d_{min} on the DPI with respect to a muscle protein structure, myC1 initially determined by MAD at 1.55 Å resolution (Govada *et al.*, 2008). In Figure 9.6(a) the DPI is plotted as a function of resolution and completeness, as completeness increases the data becomes more precise, Interestingly, ~70% complete data to 1.2 Å is just as precise as 1.3 Å data that is ~87% complete. When processing data it is worth keeping the reflections that may be present in the corners of an image! From Figure 9.6(b) we can estimate the resolution where $\sigma(x)$ is at a '3σ' level of short and long bond discrimination for carboxyl versus carboxylate Asp or Glu via the Blow/Cruickshank DPI formula, i.e. 1.02 Å. Until X-ray data of this resolution is obtained it is impossible to determine the protonation state through bond-length analysis of the X-ray structure. For a more complete explanation see Fisher *et al.*, (2008).

It should be noted that such bond distance variations upon protonation are harder to apply to histidine sidechain X-ray crystal-structure analyses (see Figure 9.7); however analysis of bond angle variations offers an alternative (Berisio *et al.*, 1999). Neutron protein crystallography with a deuterated His sidechain offers an alternative way forward at lower resolutions (Ahmed *et al.*, 2007).

The determination of bond distances and angles as well as the individual atomic displacement parameters (ADPs), to yield a full, precise and accurate 3D view of a macromolecular structure are the objectives of macromolecular crystal-structure

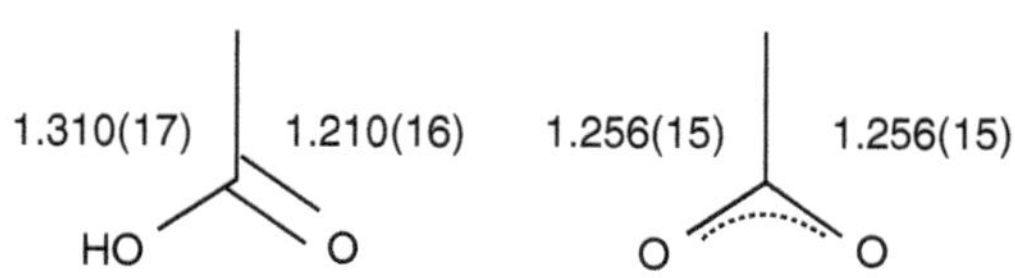

Fig. 9.5. Dictionary bond distances and spread of values (from the Cambridge Structural Database) in Å.

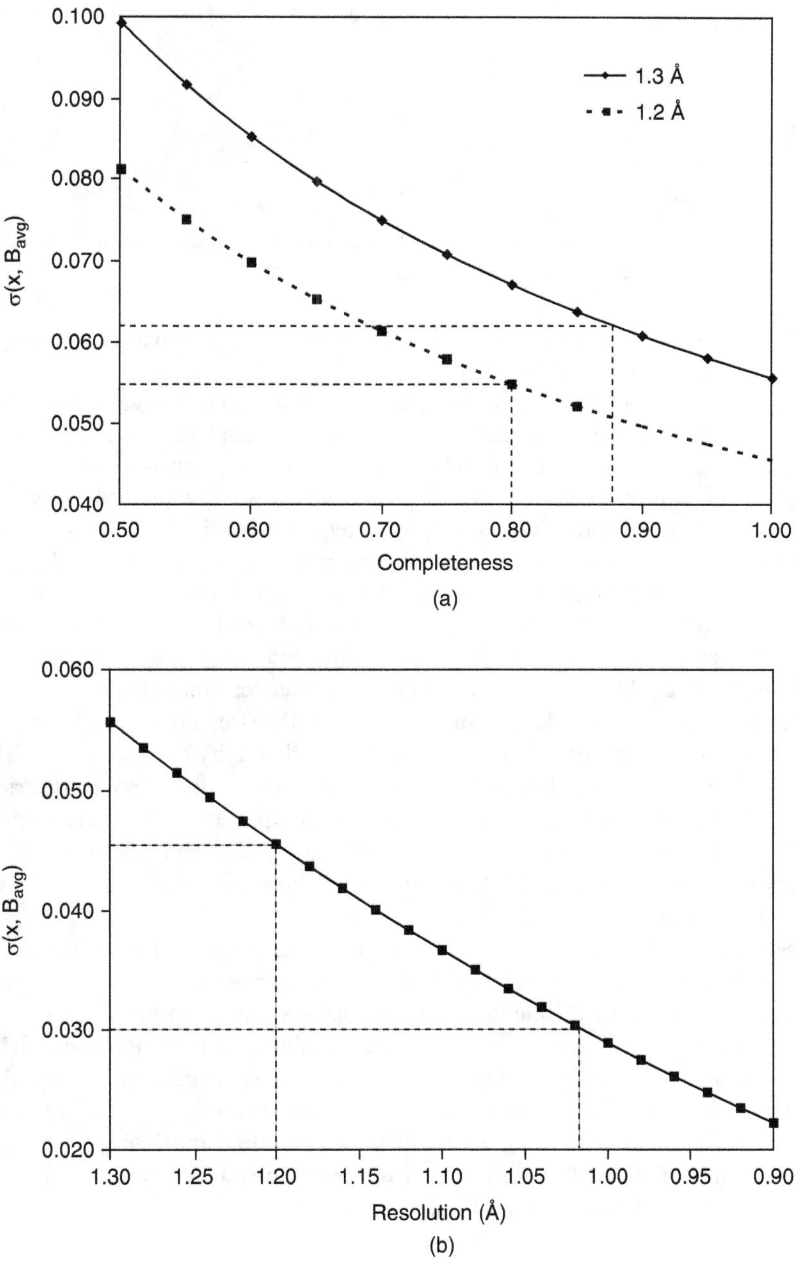

Fig. 9.6. Extrapolated $\sigma(x, B_{avg})$ versus (a) completeness at 1.2 Å and 1.3 Å and (b) resolution (for 100% completeness) for MyBP-C (Fisher *et al.*, 2008).

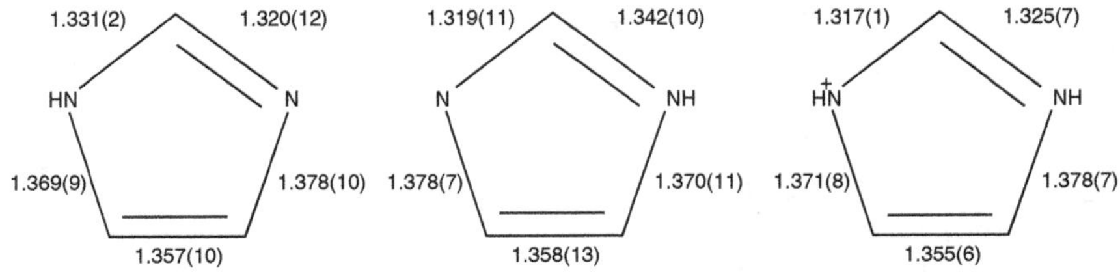

Fig. 9.7. Bond-distance variations for the protonation states in histidine (from the Cambridge Structural Database mean values and spread of values in brackets).

determination. The mobility of individual portions of a protein reduce the quality of these geometric determinations.

For the case of small-molecule (chemical) crystallography it is straightforward to judge any given value because it is given with its standard uncertainty. This is feasible because the X-ray data to refined-parameter ratio is always overdetermined usually to a factor of 10 or more even with anisotropic ADP refinement. Key quality indicators are available for a structure determination; i.e. the mean standard uncertainty of carbon–carbon bond distances, the R_{factor} and weighted R_{factor}, the diffraction data to parameter ratio and finally the temperature of the sample (room-temperature data-collection-derived structures are less precise). In the field of macromolecular crystallography the range of diffraction resolutions is considerable compared with the homogeneous situation in chemical crystallography where 0.8 Å diffraction resolution is typical. A small fraction of the structures in the Protein Data Bank are of such a quality akin to chemical crystallography. A caveat even here is needed, namely that the polypeptide loops on the protein will still not be determined from diffraction data alone and restraints must be applied. The standard uncertainties on these structural features are then in reality not easily estimated other than as deviations from idealized model geometry, in fact the same situation as applies when diffraction resolutions of 1.5 Å or poorer occur.

Restraints can be outweighed by X-ray data if the diffraction resolution is good enough. As an illustration of the impact of restraints Ahmed *et al.* (2007) undertook a study of truncations of diffraction data for concanavalin A from the experimental resolution of 0.94 Å up to 1.5 Å. A full matrix refinement inversion using SHELX (for the detailed derivations see (Rollet, 1965)) yielded standard uncertainties. As the number of X-ray diffraction data is reduced via the data truncation the restraint to being COO^- comes increasingly into play in overriding the X-ray observations. Typically, a resolution of at least 1.25 Å is needed to resolve carboxyl protonation via bond-distance analysis for well-ordered residues.

9.5 Chapter Summary

A wide variety of diffraction resolution cases exists in macromolecular crystallography each providing a different level of information content as tabulated in

Table 9.1. Like increasing the zoom on a satellite picture of the Earth, there is a coarse outline of detail seen at the lowest zoom, i.e. seeing oceans and continents, then as one further zooms in then one sees cities, and finally down to a car in a car park! So, all much like the different resolutions of a macromolecular 3D structure determination, each level of detail has a use.

10

Analysis of the molecular short-range order

10.1 Structural Data

A statistical sample of reflections, i.e. a partial or complete diffraction data set, allows the overall reflection width, the diffraction resolution at a defined signal-to-noise ($I/\sigma(I)$ usually ≥ 2), and the statistical agreement between the intensities of symmetry related reflections (termed R_{merge}) to be used as indicators of quality. R_{merge} is defined as:

$$R_{\mathrm{merge}} = \frac{\sum_{hkl}\sum_{i}|E_i(hkl) - \overline{E(hkl)}|}{\sum_{hkl}\sum_{i}E_i(hkl)} \tag{10.1}$$

where E_i is the intensity of a reflection from the (h,k,l) plane. The use of R_{merge}, diffraction resolution, signal-to-noise ratio and completeness is described in detail by Weiss (2001). The resolution at a defined overall intensity signal-to-noise ratio is dependent on the crystal quality, the beam illuminating it, the data-collection protocol and the detector properties. This measure is useful to compare when the same experimental setup and protocol are used for experiments but requires careful control of all these parameters. R_{merge} is inherently dependent on the redundancy of the data, the more a given reflection is measured the higher R_{merge}. Other statistical measures that correct for this are available, in particular the precision-indicating R-factor, R_{pim}, defined as (Weiss, 2001):

$$R_{\mathrm{pim}} = \frac{\sum_{hkl}\sqrt{\frac{N}{N-1}}\sum_{i}|E_i(hkl) - \overline{E(hkl)}|}{\sum_{hkl}\sum_{i}E_i(hkl)} \tag{10.2}$$

where N is the redundancy or multiplicity of the observed reflection.

Diffraction data collection for macromolecular crystal-structure determination is well described in various textbooks (Blow, 2002a; Drenth, 1999; Giacovazzo, 2002; Rossmann and Arnold, 2001). Specifics to synchrotron radiation are described in Helliwell (1992).

The structural solution, refinement and resulting detail seen in the electron-density maps is the ultimate measure of crystal quality. The effort needed to provide information feedback of this nature is substantial. This type of assessment is useful for understanding the crystal-growth process at an intermolecular level. In order to compare crystals of the same molecule and thereby evaluate differences in overall B_{factor} of the crystals the relative B_{factor} calculation between two crystals basically removes the complication of the molecular transform detailed shape (see Chapter 1), and therefore is calculable at less than atomic resolution. This assumes of course that the protein structure basically remains the same from crystal to crystal; in some cases protein molecule flexibility can lead to some variability. For two crystals, labelled 1 and 2, the Wilson equation can be rewritten as:

$$\frac{\langle E_{\text{obs}}(h_1, k_1, l_1)\rangle}{\langle E_{\text{obs}}(h_2, k_2, l_2)\rangle} = \mathrm{e}^{(-2(B_2-B_1)\sin^2\theta/\lambda^2)} \tag{10.3}$$

and thus a plot of log $(E_{\text{obs}}(h_1,k_1,l_1)/E_{\text{obs}}(h_2,k_2,l_2)$ has a gradient of $-2(B_2-B_1)$. This is the so-called relative Wilson plot and has been used in the analysis of a number of microgravity-grown crystals to compare them to ground-grown controls (DeLucas *et al.*, 1994; DeLucas *et al.*, 1989; DeLucas *et al.*, 1991; Koszelak *et al.*, 1995). Arai *et al.* have emphasized the use of this Wilson plot as a crystal diffraction quality indicator (Arai *et al.*, 2004). The relative Wilson plot thus normalizes the data against a reference crystal. The quality of any crystal of the same type as the reference crystal is given by the sign and the magnitude of the gradient of the slope from the resulting relative Wilson plot. This is a useful technique for comparison within a given population of crystals but again, as in the case of resolution, requires careful control of all the experimental parameters. For identical crystals the slope is zero, if one crystal is improved or degraded over the other the slope is non-zero.

10.2 Thermal Motion and Diffuse Scattering

What is diffuse scattering? Basically, not all the diffracted X-ray photons from crystals end up in the Bragg reflections from specified (h,k,l) planes. Indeed, for quite a large number of macromolecular crystals, the non-Bragg diffraction or diffuse scattering as it is known, is strong in intensity. The diffuse scattering is due to a breakdown in the periodicity of the crystal and carries information on the mobility and flexibility of the molecules in the crystal (Welberry, 2004). It may arise from several sources including:

- thermal diffuse scattering;
- static disorder scattering;
- solvent disorder.

The static or dynamic displacement of atoms in crystals causes a breakdown of translational symmetry of the crystal, leading to a reduction in the Bragg intensities

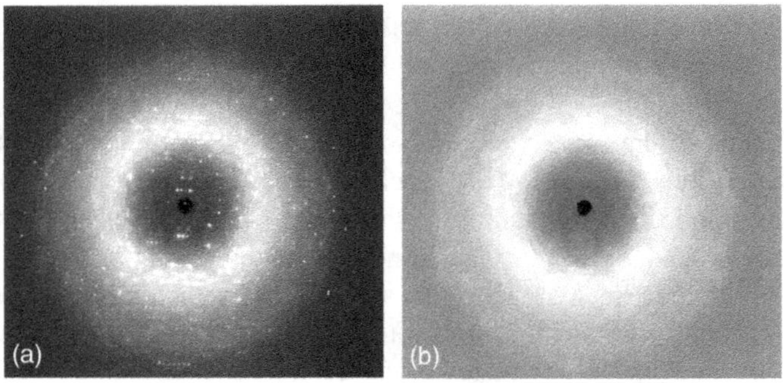

Fig. 10.1. (a) Raw image of a room-temperature, 5-s still exposure (roughly 1.5 Å resolution at the edge) from a crystal of Staphylococcal nuclease. Note the structure in the diffuse scattering observable in the image. The authors note that the distribution of diffuse features is consistent with the unit-cell space group $P4_1$. (b) Diffraction image shown after polarization correction, solid-angle normalization, and mode filtering to remove Bragg peaks. From Wall *et al.* (1997) with permission of the authors and PNAS.

at high resolution and the appearance of diffuse scattering at and between the reciprocal lattice positions. In the case of macromolecular crystals, diffuse scattering is often quite strong, can be rich in detail and apparently distinctive to a specific macromolecule and/or crystal. It represents a potentially valuable source of information regarding atomic displacements, for example a plausible hinge bending motion in 6-phosphogluconate dehydrogenase (Helliwell *et al.*, 1986a). A comprehensive review of the topic is given by Moss *et al.* (2003).

A recent example is that of Wall *et al.*, (1997) who have developed methods for obtaining and characterizing three-dimensional maps of the reciprocal-space distribution of diffuse X-ray scattering from protein crystals, and have used the methods to study the nature of disorder in crystals of *Staphylococcal* nuclease, Figure 10.1.

Static disorder arises when unit cells exist with different arrangements of the time-averaged positions. Static orientational disorder occurs in molecular crystals where molecules, flexible domains, or side groups take up different orientations, breaking the translational symmetry.

Dynamic disorder arises from thermal vibrations and is present in all crystals. Two types of lattice vibrations may be distinguished, acoustic modes due to the propagation of ultrasonic waves in the crystal and optic modes of vibration such as are observed in infra-red and Raman spectra. Ultrasonic vibrations give rise to thermal diffuse scattering, which peaks primarily at the reciprocal lattice positions and is observed characteristically as a feature at and around the Bragg peak. Optic mode vibrations along with other disorder modes give rise to diffuse scattering, which is distributed continuously but non-uniformly throughout reciprocal space. Thus, ultrasonic effects are long range encompassing many unit cells, and thus yielding X-ray diffuse peaks centred at the Bragg reflection positions. The optic modes are involved in short-range order and yield broadly dispersed diffuse diffraction.

Many interesting examples of diffuse, mainly X-ray, scattering are logged in the macromolecular, and small-molecule, crystallography literature. The topic is large and too extensive to review here. A survey of the types of diffuse scattering seen in macromolecular crystals is given by Glover *et al.,* (1991). The X-ray diffuse scattering from protein crystals is relatively very weak generally and it is largely with the advent of synchrotron radiation that it has become feasible to study this (Wilson *et al.*, 1983).

10.2.1 Diffuse scattering as a source of measurement error in the Bragg intensities

As macromolecular model refinement in crystallography is based only on the Bragg reflection intensities, any other intensity contribution to a spot intensity besides the Bragg reflection will be a source of error. A useful example can be taken from the electron crystallography study of bacteriorhodopsin (bR) crystals (Grigorieff and Henderson, 1995). Diffuse background was significant and in certain regions of the diffraction pattern even resulted in systematically negative intensity measurements. Removing the diffuse background (based on models of its potential cause) produced more accurate measurement of structure-factor amplitudes. The best correction for the several models tested was obtained by assuming that the disorder came from random displacement of the trimer of bR molecules as a rigid unit. A correction was applied to the intensity measurements and the number of intensities previously measured wrongly to be negative (about 10%) was reduced by a factor 4-fold; in agreement with that expected from counting statistics. On average, the correction had a magnitude of 18% of the original intensities, although the improvement in the data, when used for refinement of an atomic model, lead to a marginal R-factor reduction of 0.40%. In X-ray and neutron macromolecular crystallography such diffuse scattering corrections are rarely made, if ever.

The extent to which diffuse scattering from macromolecular crystals can be measured and harnessed is certainly recognized and new computational software 'tools' have been developed as exemplified by Wall *et al.,* (1997) for example. The potential for systematic crystal growth avenues to be explored with a given macromolecule can surely benefit from:

(i) robotics to aid systematic exploration of crystal forms and achieve more reproducible growth conditions; this might also include systematic introduction of impurities to stimulate diffuse scattering effects as a learning exercise;
(ii) molecular-dynamics simulations of ever increasing time sequences, e.g. now as long as 10 ns from multiple starting points to evaluate interdomain motions and 'plasticity' of proteins; see Hery, *et al.,* (1998)
(iii) a synergistic interaction with small-molecule crystal diffuse scattering measurement and software tools.

In addition, the correction, i.e. subtraction, of diffuse scattering contributions to the spot intensities to better extract the Bragg reflection intensities should be investigated systematically to assess potential improvement in protein model refinements.

10.3 Chapter Summary

Whilst the transition from short-range order (this chapter) to long-range order (the next chapter) is not a sharp transition, i.e. there is a steady progression, there are metrics to judge the shorter-range effects. Not least the overall flexibility of the molecular motif can be described by the average B_{factor} of the diffraction data. This relates closely to the crystal's overall B_{factor}. The latter is, however, affected by experimental conditions for the measurement of the diffraction data. Thus, comparisons of crystal-growth conditions and resulting crystals should be under as identical experimental conditions for measuring the diffraction data as possible.

Diffuse scattering is of two types; one that is associated with short-range order and one that is associated with long-range order. For convenience, and the special nature of the topic, we have kept these two aspects in this chapter.

11

Analysis of long-range order

Long-range order and deviations from perfection requires detailed analysis at and in the immediate vicinity of Bragg reflections. Diffraction measurements are the optimum measure of crystal quality whatever a crystal may look like visually. There are several types of physical X-ray diffraction analyses now employed in addition to standard diffraction data collection for structural studies, e.g. reflection profiling, topography and reciprocal-space mapping (Boggon *et al.*, 2000; Snell *et al.*, 2003). The analysis techniques can be divided into properties that are measured from single reflections, a statistical sample of reflections or as complete a data set as possible. Reflection profiling can be carried out similar to standard structural data collection with the only difference being the minimization of instrument and beam contributions and use of a fine-step crystal rotation (Bellamy *et al.*, 2000; Boggon *et al.*, 2000; Colapietro *et al.*, 1992; Fourme *et al.*, 1995; Fourme *et al.*, 1999; Snell *et al.*, 1995b). At the synchrotron the spectral parameters are controlled by the monochromator and the X-ray beam divergences are controllable via precision slits in the vertical and horizontal planes, finally limited by the intrinsic properties of the synchrotron machine electron beam emittance. Reflection profiling can also be accommodated in the laboratory, at the expense of flux, using a Bartels 4-bounce monochromator (Bartels, 1983; Hu *et al.*, 2001). Topography uses a similar optics setup but with a spatially sensitive detector, e.g. photographic film, so that the detail in the reflection can be resolved (Boggon *et al.*, 2000; Dobrianov *et al.*, 1998; Fourme *et al.*, 1999; Otalora *et al.*, 1999; Stojanoff and Siddons, 1996; Stojanoff *et al.*, 1997; Stojanoff *et al.*, 1996). Finally, in reciprocal-space mapping, an additional crystal is inserted between the sample and the detector (called an analyser crystal) allowing the reciprocal lattice to be resolved in two dimensions (Boggon *et al.*, 2000; Fewster, 1997; Volz and Matyi, 1999; Volz and Matyi, 2000; Volz and Matyi, 2001).

11.1 Reflection Profiling

Shaikevitch and Kam (1981) published one of the first studies on the use of reflection profiling as an indicator of macromolecular crystal perfection. Subsequently, Helliwell and co-workers made use of the synchrotron-radiation properties described previously to minimize the geometric and spectral contributions of the X-ray source to the experimental data (Colapietro *et al.*, 1992; Helliwell, 1988). The measurements of mosaicity were made initially by recording reflections individually with a scintillation counter mounted in the equatorial (vertical) plane and by rotating the crystal about a horizontal axis (Colapietro *et al.*, 1992; Fourme *et al.*, 1995; Helliwell, 1988; Ng, 1997; Snell *et al.*, 1995b). This experimental setup minimized the Lorentz effect and essentially eliminated the contribution from the horizontal divergence of the synchrotron beam. Mosaicity analysis of chicken egg-white lysozyme, apocrustacyanin-C_1, and thaumatin crystals established a physical basis for the improvements seen in these microgravity-grown samples. The reduction in the mosaic spread in the microgravity-grown crystals produced a corresponding increase in the signal-to-noise ratio of the reflection. The minimum mosaicities recorded were 0.005° for lysozyme, 0.030° for apocrustacyanin C_1, and 0.018° for thaumatin (Ng, 1997; Snell *et al.*, 1997; Snell *et al.*, 1995b).

Laue diffraction was also used with long distances between crystal and detector (~1 m) (Snell, 1995; Snell *et al.*, 1995a); see below. Monochromatic methods initially looked at a few, low-resolution reflections recorded one at a time (Boggon *et al.*, 2000; Colapietro *et al.*, 1992; Fourme *et al.*, 1995; Snell *et al.*, 1995b). Later methods were developed using area detectors (Bellamy *et al.*, 2000; Ferrer and Roth, 1998) to analyse hundreds of reflections in a short time (Bellamy *et al.*, 2000; Lovelace *et al.*, 2000).

A single reflection can provide an indication of the crystal mosaicity. It is most useful when comparing identical reflections in a sample population of crystals as anisotropic effects may be present. At a minimum, reflections at or near to identical Bragg angles must be compared. Implicitly a reduction in mosaicity implies an improved spatial coherence length of a crystal's mosaic block and leads to an increase in signal-to-noise for identical reflections from otherwise identical crystals. The analysis of individual reflections can provide a measure of the long-range order within the crystal. In addition, by making measurements in multiple regions of reciprocal space, crystal anisotropy can be investigated.

In the Ewald sphere construction, the sphere is an infinitesimally thin shell. This corresponds to a perfectly monochromatic beam with no angular divergence and the reflection width is purely governed by the mosaicity, Figure 11.1(a). A beam with non-zero beam divergence, Figure 11.1(b), and finite bandwidth, Figure 11.1(c), can be modelled by Ewald spheres with finite shell thicknesses. A perfect crystal would have extremely small, almost infinitesimal, reciprocal-lattice points. However, the mosaicity of a real crystal broadens the reciprocal-lattice points into finite volumes. If the reciprocal-lattice point can be totally encompassed in the thickness of the shell of the Ewald sphere, then the effect of the crystal quality on the reflection parameters will be masked and, in effect, only the beam parameters will be measured. When investigating crystal quality, the probe, the X-ray beam, has to be carefully configured to prevent this. Typically, in ordinary data collection the beam is focused to increase flux on the crystal. At synchrotron beamlines the bandwidth is not as narrow as it could be for the same reason.

An additional effect is that the intensity of the reflections (reciprocal-lattice points) varies through the Ewald sphere at differing angles of incidence to the surface of the

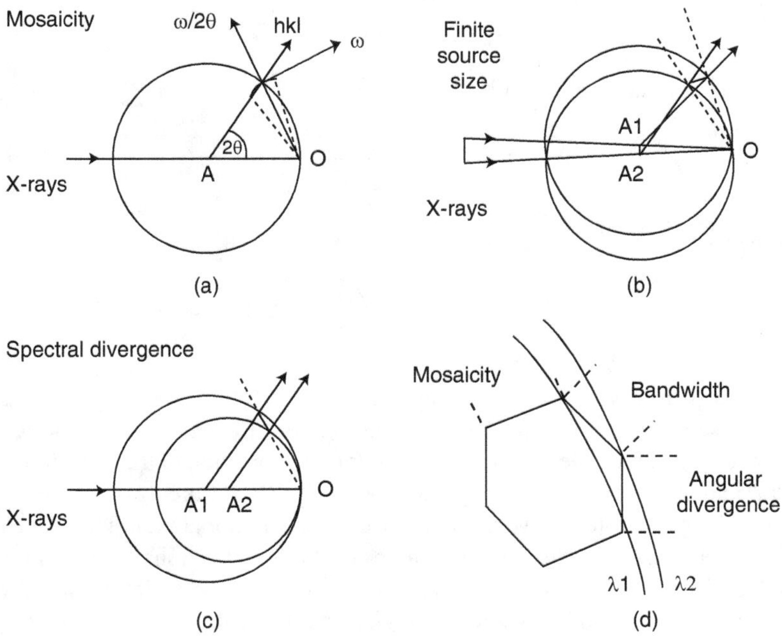

Fig. 11.1. Ewald-sphere construction showing the influence of crystal mosaicity and experimental parameters on reflection broadening in reciprocal space (from Snell *et al.* (2003) with permission of Elsevier). The effect of mosaicity, geometric divergence and spectral divergence are shown in (a–c). The overall influence of these effects (simplified) is illustrated by an enlargement of a reciprocal-lattice point in (d).

sphere. This, of course, is the Lorentz effect and causes the interaction width of the reflection to be increased independently of the quality of the crystal or the characteristics of the incident beam.

The angular width for a reflection is given by (Bellamy *et al.*, 2000; Greenhough and Helliwell, 1982a; Greenhough and Helliwell, 1982b; Helliwell, 1992);

$$|\phi_{\mathrm{R}}| = \sqrt{L^2\zeta^2\gamma_H^2 + \gamma_V^2} + \frac{L}{d}\cos\theta_{hkl}\left[\eta + \left(\frac{\delta\lambda}{\lambda}\right)\tan\theta_{hkl}\right] \tag{11.1}$$

Here, ϕ_{R} is the measured reflection width, ζ is the position of the corresponding reciprocal-lattice point projected onto the rotation axis, d is the resolution ($d=\lambda/2\sin\theta_{hkl}$), η is the mosaic spread and L is the correction for the Lorentz effect. Because a horizontal rotation axis is generally used at synchrotron beamlines we use H and V to denote directions along the rotation axis and perpendicular to both the rotation axis and the beam respectively, γ_V and γ_H are the vertical and horizontal angular divergence at the sample. ζ^2 is given by;

$$\zeta^2 = \left(\frac{H^2}{H^2 + V^2}\right) \sin^2(2\theta_{hkl}) \tag{11.2}$$

The Lorentz correction is given by;

$$L = \frac{1}{\sqrt{\sin^2(2\theta_{hkl}) - \zeta^2}} \tag{11.3}$$

The reflection angle $2\theta_{hkl}$, is given by;

$$2\theta_{hkl} = \tan^{-1}\left(\frac{\sqrt{H^2 + V^2}}{\mathrm{XTD}}\right) \tag{11.4}$$

where XTD is the crystal to detector distance.

One can see that γ_V broadens the reflections universally over the detector, whereas the effect of γ_H on the reflection width depends on the position of the reflection on the detector and is a maximum along the horizontal. The Lorentz effect is always maximal along the rotation axis, which in this case is horizontal. The wavelength dispersion term has its largest effect on high-resolution reflections. In eqn (11.1), the correlated dispersion is ignored. Correlated dispersion is the variation of the wavelength across the beam convergence angle, and is negligible with X-ray optics suitable for reflection analysis. Accurate structural and crystal-quality data collection has to overcome or correct for these contributions to the reflection profile in the integration process.

An alternative approach for measuring crystal perfection is the Laue method that uses polychromatic ('white' – full bandpass or 'pink' – partial bandpass) incident radiation to illuminate a stationary crystal (Snell *et al.*, 1995b). The method is sensitive to the mosaicity of crystals and simultaneously records a large number of reflections. Like the monochromatic method, Laue experiments require a highly parallel incident beam if they are truly to examine the intrinsic properties of the crystal. In the Laue case the mosaicity, η, is derived from the radial extension, Δ_{radial}, of the reflections (Cruickshank *et al.*, 1992);

$$\Delta_{\mathrm{radial}} = 2\eta \frac{\mathrm{XTD}}{\cos^2 2\theta} \tag{11.5}$$

This assumes an incident beam of zero divergence, and the relationship becomes more complicated if that criterion is not met. A large crystal-to-film distance and a fine pixel-size detector, e.g. X-ray film, are required to make accurate measurements of Δ_{radial} (Helliwell *et al.*, 1996; Snell *et al.*, 1995b; Weisgerber and Helliwell, 1993a).

Home laboratory sources with focusing mirrors or graphite monochromators are not suitable instruments to study typical macromolecular crystal quality because of their high beam divergence. Standard data-processing packages output what are described as mosaicity values but these are a convolution of parameters and do not provide a true indication of crystal quality. The home source can be configured for accurate crystal-quality measurements but only at the expense of X-ray intensity. For example, a Bartels-type (Bartels, 1983) monochromator can be used to condition the beam. This type of monochromator can achieve a very fine geometric divergence of 52 μrad and a spectral divergence of 1.5×10^{-4} using the Ge(220) reflection. Other optical systems, e.g. parabolic graded mirrors can achieve reductions in the divergence characteristics (Volz and Matyi, 2000) while increasing the available flux but do not approach that available from the synchrotron.

Evaluating a statistically valid sample of indexed reflections becomes very important for comparative studies involving many crystals, for example, crystals grown by different methods, crystals of different morphologies, or for comparing crystal manipulations such as cryo-cooling protocols. The CCD-based techniques prove ideal for this and were first applied to crystals of *E. coli* manganese superoxide dismutase (MnSOD) (Borgstahl *et al.*, 2000). In a total of one degree of data, the mosaicities of 260 reflections were measured. The mosaicity averaged 0.010° (*s.d.* 0.004°), measured as the FWHM, and ranged from 0.001° to 0.019°. Each reflection could be fitted with two Gaussian curves indicating that the crystal was composed of at least two mosaic domains. Indexed reflections proved critical and allowed the anisotropic mosaicity to be related to the crystal packing based on the work of Ferrer and Roth (1998). Another study on lysozyme (Snell, 2001) developed a general expression;

$$\eta_{hkl}^{\mathrm{calc}} = \frac{\eta_{abc}\left(\frac{(ah)^2+(bk)^2+(cl)^2}{a^2+b^2+c^2}\right) + \eta_{def}\left(\frac{(dh)^2+(ek)^2+(fl)^2}{d^2+e^2+f^2}\right) + \eta_{mno}\left(\frac{(mh)^2+(nk)^2+(ol)^2}{m^2+n^2+o^2}\right)}{h^2+k^2+l^2} + \eta_{\mathrm{const}} \quad (11.6)$$

where (a,b,c), (d,e,f) and (m,n,o) are real space vectors in the crystal lattice coordinate system, h, k and l are the refection indices and η_{const} is the isotropic component of the mosaicity. Lysozyme proved to be isotropic in terms of mosaicity but this equation allows anisotropic mosaicity to be probed in terms of any defined direction, e.g. one related to the lattice or to the surface morphology.

Another example was a comparison of insulin crystals grown on earth with those grown in microgravity (Borgstahl *et al.*, 2001). Using fine ϕ sliced data (0.001° steps), between 447 and 502 reflections were profiled for each of six microgravity-grown insulin crystals at the Stanford Synchrotron Radiation Laboratory (SSRL) beamline 1–5. The contribution of the instrument to the reflection profiles was limited to a broadening of 0.0016° minimally. The best microgravity crystals had an average η of 0.002° with a standard deviation of only 0.001° – near the limit of resolution of the instrument configuration used. Two of the earth crystals had fairly low mosaicity with average η values of 0.013° (s.d. 0.004°) and 0.017° (s.d. 0.005°), respectively, yet these η values were 6.5 and 8.5 times higher than the best microgravity-grown crystals and both earth-grown crystals were relatively poor diffractors. For any given earth-grown crystal, the η values for individual reflections varied over a surprisingly large range, with standard deviations of 0.004 to 0.024°. The spread in η for microgravity crystals was 4–5-fold narrower with standard deviations ranging from 0.001 to 0.005°. In a few cases, the best earth η values overlap the worst microgravity values. This illustrates the importance of collecting a statistically significant number of reflections from each sample since an unlucky selection of a few reflections or just a couple of crystals in each population could lead to an erroneous conclusion.

It is important, not only to collect a statistically significant number of reflections, but also to collect data from multiple samples: in this case six crystals of each kind. The microgravity crystals were on average 34 times larger in volume, had 7 times lower mosaicity, had 54 times higher reflection peak heights and diffracted to significantly higher resolution than their earth-grown counterparts. Nevertheless, crystals with reduced mosaicity produced data with a higher signal-to-noise ratio.

During structural data collection setting the correct ϕ step can take advantage of reduced mosaicity to maximize the signal-to-noise ratio, thereby improving the useful resolution in the data (Pflugrath, 1999). Reduced mosaicity increases the number of fully recorded reflections per image, and reduces reflection–reflection spatial overlap (Fourme *et al.*, 1999). Fine-sliced images using oscillation methods can be used to take advantage of low mosaicity but the method does present technical challenges. The data may suffer from increased detector readout noise, and the shorter exposure, narrower angular range images place more stringent requirements on the hardware for shutter timing and goniometer control (Pflugrath, 1999). The time lost during detector readout is also increased. In studying mosaicity, superfine ϕ slicing provides the necessary detail. However, for structural data collection where the beam is not as parallel and possibly not as monochromatic, there is little or nothing to be gained with oscillations less than one third to one half of the greater of the beam contribution or the crystal mosaicity (Pflugrath, 1999). It is important, of course, to understand the characteristics of the beamline before starting data collection, and to process the data as they are collected to maximize their quality. With a new generation of fast detectors and shutterless data collection, very low mosaicity will be properly and routinely harnessable in room-temperature data collection.

11.2 Topography

A topograph is a detailed image of those parts of the crystal contributing to the reflection at a single crystal position (Stojanoff and Siddons, 1996; Stojanoff *et al.*, 1997). For a perfect crystal the whole crystal contributes to the Bragg reflection over an infinitely small Bragg angle. But no crystal is perfect, and typically, for a high-quality crystal, distinct regions of the crystal produce contrast at slightly differing Bragg angles. For poor-quality crystals the topograph appears uniformly grey, the contribution of different parts of the crystal at different Bragg angles are smeared out.

X-ray topography is the study of ways that irregularities in the lattice cause locally changing diffracted intensities (contrast) within individual reflections (for protein crystal samples see Boggon *et al.* (2000)). Topographs are a measure of the scattering power of a crystal as a function of position across the diffracted X-ray beam. In most cases, it is not the defect itself but the variations in the lattice surrounding the defect that produces the contrast. Intensity variations are related to the type and volume distribution of defects. Three causes of contrast are: orientation variations due to domain misalignment, extinction caused by a high strain gradient, and dynamical scattering effects. The last is small for weakly scattering macromolecules. A high-quality region of the crystal will have a uniform dark or light area in the topograph. The maximum spatial resolution obtainable in a topograph is about 2–3 μm with photographic film and 1 μm with nuclear-emulsion plates. Recently, CCD based imaging has been used to examine macromolecular crystals at a spatial resolution of $\sim$8 μm (Lovelace *et al.*, 2004).

11.2.1 Photography-based topography

Topography on macromolecular crystals was suggested by Shaikevitch and Kam (1981). Stojanoff and Siddons (1996) used the white-beam Laue method to study lysozyme crystals. Highly strained regions, high densities of defects, and quite perfect regions were seen. The topographs were surprisingly detailed. Fourme *et al.* (1995) used reflection profiles to take topographs at different Bragg angles of multiple peaks seen in the same reflection, again from lysozyme. They discovered separate regions or domains of the crystal contributing to each peak of the total reflection.

Topography has been used as an effective technique to study the effect of solution variations during crystal growth (Dobrianov *et al.*, 1998). Topographs of lysozyme crystals subjected to deliberate variation of temperature, pH, or mother liquor concentrations during their growth revealed several general effects. Lysozyme is relatively insensitive to changes in growth conditions compared to most macromolecules, so the changes employed were comparatively large. Temperature was varied from 295 to 288 K, pH from 4 to 5 and in combination protein concentration reduced from 65 to 11 mg ml^{-1}, while salt was increased from 0.45 to 1.2 M. The authors also studied the effects of protein concentration by transferring growing crystals from a 27 to a 41 mg ml^{-1} protein concentration solution. This increase in protein concentration mimics a seeding experiment. In crystals subjected to a pH change, the scattered intensity from the boundary layer just outside the pre-change region differs strongly from both earlier and subsequent regions. The lattice growing during the change is more disordered than that before and that shortly after. It seems that crystal perfection recovers in subsequent lattice growth. A similar effect is seen for concentration changes of both the protein and salt. Temperature change causes a difference in the mosaicity or unit cell dimensions. A factor of 3 increase in the growth rate did not produce substantial features in the resulting topographs. This suggests that in the growth process, a change of protein concentration in the drop does not affect the quality of the resulting crystal. By application of reflection profiling in the same experiment it was concluded that the contrast variation seen in the topograph is primarily due to spatial coherence length, i.e. the length over which a number of unit cells align to form one mosaic domain.

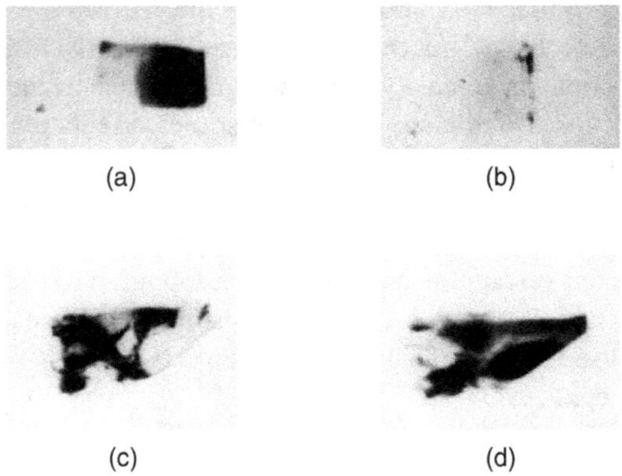

Fig. 11.2. Topographs from (a) and (b) a microgravity grown lysozyme crystal, 1.1×0.9 mm in projection, and (c) shown further enlarged, an earth-grown lysozyme crystal 0.8×0.6 mm in projection (Boggon *et al.*, 2000) used with the permission of the authors and the IUCr.

Topographs acquired at successive angles within the reflection profile will map out the contribution of the crystal to each point of that profile. Figure 11.2 illustrates topographs from two high-quality lysozyme crystals. In 11.2(a) and 11.2(b) the crystal clearly consists of two major domains, whereas the crystal illustrated in (c) and (d) consists of several domains separated by boundary areas (Boggon *et al.*, 2000; Snell *et al.*, 2003). With an undulator source the angular divergence of the beam can be very small and the spatial resolution in the topograph high. The different growth sectors within the crystal can be imaged, and more remarkably, fringes at the boundaries of those growth sectors can be seen. Topography provides a strong but qualitative method suited to the study of crystal growth and other practical applications such as the study of cryo-protectant effects on cooling (Kriminski *et al.*, 2002).

11.2.2 Digital-based topography

Photographic techniques, either film or nuclear emulsion plates, provide a very small pixel size for imaging small details in the reflection. Drawbacks to the use of film or emulsion plates include slow processing time, variation between batches and processing chemicals, limited dynamic range, and the extra digitization step. Repeatable high-accuracy quantitative analysis with these techniques is difficult and time consuming. Film and, to a lesser extent, emulsion-plate processes are slowly being replaced with digital equivalents in many fields and with the pixel size of CCD devices decreasing it has been possible to use CCD-based technology to digitally acquire topographic images (Lovelace *et al.*, 2004).

11.3 Reciprocal-Space Mapping

The mosaicity measured from the crystal is a global measure of several effects. Simple rocking-curve measurements cannot separate effects arising from strain (i.e. lattice-spacing variations) from those arising from the classical definition of mosaicity that involves only domain misalignment. The separation requires that reciprocal space is mapped in two dimensions. Effects in crystal space broaden the lattice point in reciprocal space, Figure 11.3.

The technique used for this purpose is the triple-axis spectrometer first used for neutron scattering (Brockhouse, 1955). The application of reciprocal-space mapping to X-ray techniques is well described elsewhere (Fewster, 1996; Fewster, 1997). Boggon *et al.* (2000) combined this with mosaicity and topography methods to look at microgravity and ground-grown crystals. Figure 11.4(a) shows the instrumental geometry used and an example of a reciprocal-space map from a microgravity-grown lysozyme crystal.

Although the term reciprocal-space mapping can be used to describe all methods of diffraction data collection (Fewster, 1997) we use it in a more limited sense to describe examining a volume of reciprocal space in two or three dimensions. Both reflection profiling and topography image the reciprocal lattice over a relatively large volume, causing much information about the shape of the reciprocal-lattice point to be lost. Reciprocal-space mapping provides the shape information, details lost from the other techniques. The effects shown in Figure 11.3 contribute to the measured mosaicity. Reciprocal-space mapping allows us to understand mosaicity in terms of the components that contribute to it.

Reciprocal-space mapping is accomplished by sampling the reflection profile using an analyser crystal in the path of the diffracted beam, Figure 11.4(a). Reciprocal-space mapping is a high-fidelity technique, but because it is time consuming only a few reflections can be studied from one sample. Therefore, the most effective study would combine reciprocal-space mapping with one or more other techniques (Boggon *et al.*, 2000). For example, one would identify reflections of interest by area detector mosaicity analysis and then these selected reflections could be studied in detail by reciprocal-space mapping. Reciprocal-space mapping of macromolecular crystals was first performed in the laboratory using a Bartels monochromator system. Lysozyme (a well-behaved and frequently used model macromolecular crystal sample) scattered weakly but produced very sharp profiles (Fewster, 1997). Experiments with synchrotron radiation produced similar results with improved signal-to-noise (Boggon *et al.*, 2000). By recording maps at multiple χ positions, i.e. a rotation parallel to the beam, a three-dimensional profile of the reciprocal lattice can be built up (Fewster, 1997). Lysozyme crystals were found to present a complex analysis problem since reciprocal-space mapping data reveals that they appear to lie at the convergence of the kinematical (ideally imperfect crystal model) and dynamical (ideally perfect crystal model) treatments of diffraction (Volz and Matyi, 2000). Kinematical diffraction ignores the interaction of wave fields within the crystal and is valid for a crystal that is small compared to the extinction distance ε defined by Bowen and Tanner (1998);

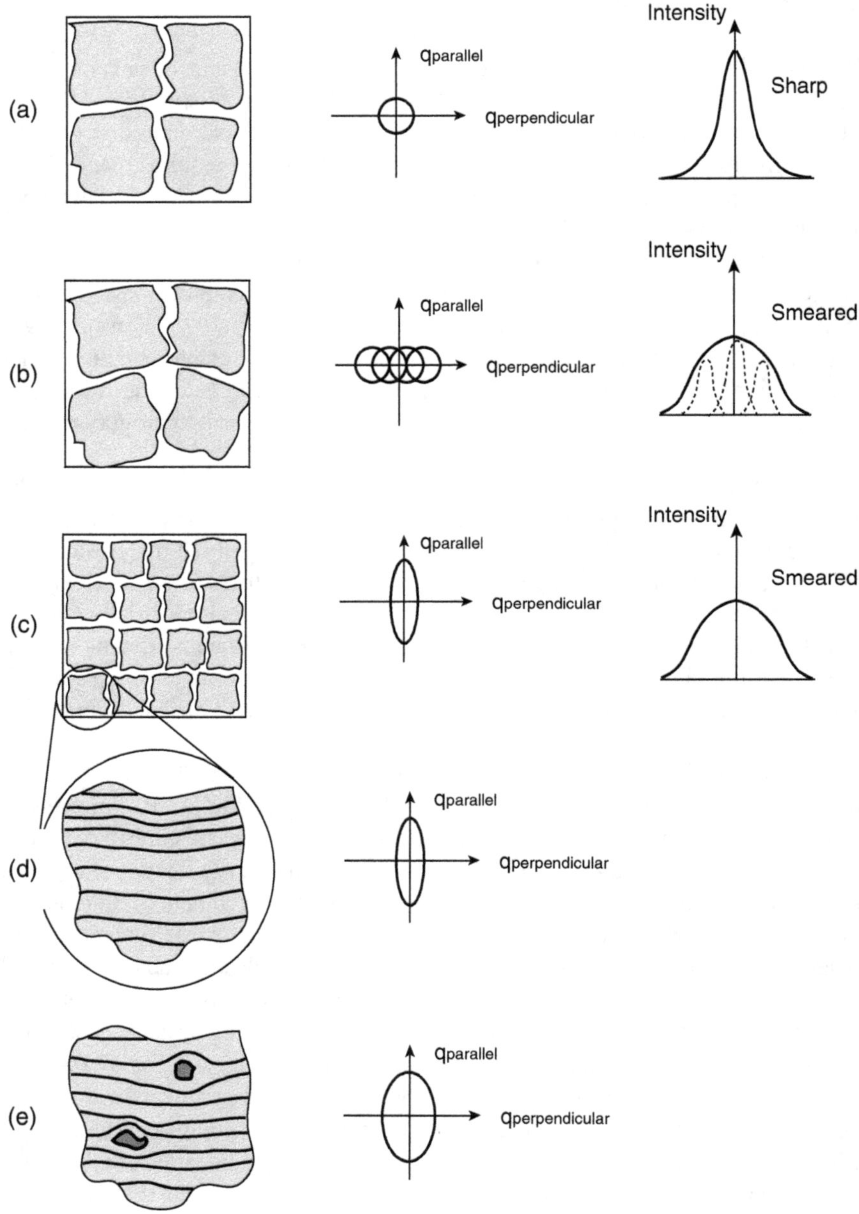

Fig. 11.3. Schematic illustration of crystal properties in the first column with the resulting reciprocal-space map (second column) and the effect on the reflection profile (third column). Shown are examples where (a) several well-aligned domains are present (b) these domains are misaligned or (c) reduced in volume. Domains within the crystal (shown enlarged in (d) and (e)) can have variation in the d spacing resulting from effects such as (d) cryo-cooling and/or (e) impurities. A typical crystal consists of multiple domains. From Snell *et al.* (2003) with permission of Elsevier.

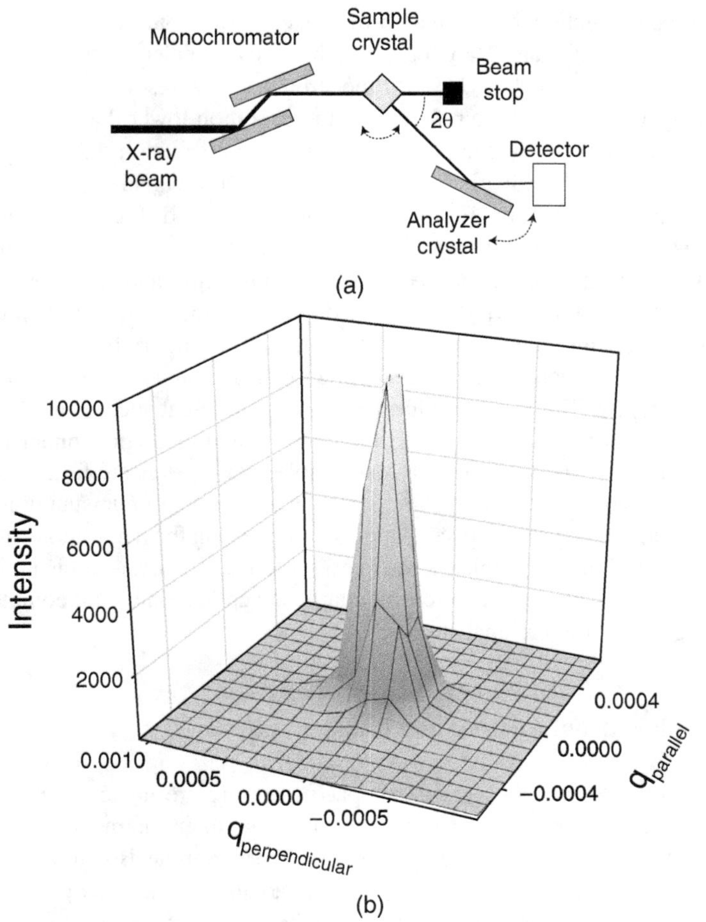

Fig. 11.4. Reciprocal-space mapping showing (a) instrumentation, (b) the resulting reciprocal-space map.

$$\varepsilon = V_o/(r_0 C|F_h|\lambda) \tag{11.7}$$

where C is the polarization factor, V_o the volume of the unit cell, r_0 the classical electron radius, $|F_h|$ the amplitude of the structure factor and λ the wavelength. Dynamical theory allows the coupling of the wave fields within the crystal and accounts for extinction effects. For X-ray wavelengths and macromolecular crystals the extinction distance has been reported to be on the order of a mm (Fourme *et al.*, 1995; Helliwell, 1988; Volz and Matyi, 2000). The mosaicity can be predicted from both kinematical and dynamical theory. The values predicted from both theories turn out to be similar (Volz and Matyi, 2000). Dynamical theory can have an important impact for structural crystallography on the accuracy of the integrated intensities especially of the lower resolution more intense reflections. Polykarpov and Sawyer (1995) derived an extinction correction that takes into account dynamical properties in macromolecular crystals. They found that in the case of alcohol dehydrogenase the correction may be as much as 15% for the strongest, low-resolution reflections, and that as many as 20% of all the reflections at a resolution lower than 3.4 Å had to be corrected by more than 2% compared to kinematical diffraction data.

The considerable length of time required for reciprocal-space mapping makes radiation damage a concern. Fortunately, when unfocused, highly monochromatic radiation is used, samples receive far lower doses than for an equivalent time of structural data collection. Radiation damage is both time and dose dependent but Volz and Matyi (2001) report a case of five days of continuous radiation not affecting data from a lysozyme sample on a well-conditioned laboratory beam.

Reciprocal-space mapping reveals information that cannot be seen through mosaicity or topography. The technique has been used with great success in the semiconductor industry owing to a comprehensive practical and theoretical understanding of the sample material (Fewster, 1997). Macromolecular crystals are far more complex systems, and theory has yet to catch up with experiment in understanding just how much information reciprocal-space mapping can reveal in the macromolecular crystal world. It is one of the developing areas in crystal-quality analysis but is unlikely ever to be used as a routine method before standard data collection would commence on a given sample for structure analysis.

11.4 Combinational Analysis and Chapter Summary

Mosaicity, topography, and reciprocal-space mapping are all techniques to probe the physical characteristics of the crystals through their interaction with X-rays. The techniques described are complementary. For example Boggon *et al.* (2000) combined the three techniques with synchrotron radiation in the study of microgravity- and ground-grown crystals. Only a small number of samples were used, but microgravity crystals showed a reduced mosaicity. Reciprocal-space maps saw no change in stress, and topography showed that the majority of the crystal was contributing to the peak of the reflection at the appropriate Bragg angle in the microgravity case. Each technique provided unique and complementary information.

In terms of structural crystallography, i.e. solving and understanding the structure of a macromolecule of interest, having a high-quality crystal is clearly desirable. The techniques described here are not part of routine data collection. Of the techniques described, mosaicity measurements can be performed relatively easily and are the most useful in the short term. The quality of the data can be optimized by matching the oscillation range to the mosaicity. The background noise in an oscillation image builds up throughout the oscillation range but the reflection is only recorded over a finite angle. With the advent of faster detectors, 'ideal' data collection is now possible by continuous rotation with real-time detector readout offering effectively infinitely fine slicing.

Mosaicity, topography, and reciprocal-space mapping are diagnostic techniques that allow us to ask questions about the practical effects of the crystal-growth process and the data-collection practices in order to optimize them. They offer quantitative data about crystal-growth methods, biochemical properties, and practical matters such as cryo-cooling protocols, cryogens, and crystal handling for automated studies. The resolution of the structural data and corresponding electron-density maps provide us with an indication of the short-range quality. The techniques described here give us a measure of long-range order. Many of the crystal-quality techniques have been developed with lysozyme, the future will see them being applied to either cases. Eventually, crystal growth, now an empirical process of rational trial and error guided by past experience, may be understood in far greater detail with information from reflection analysis. A surprise has been just how ordered macromolecular crystals can be. This offers potential in new phasing methods such as multiple-beam diffraction (Weckert and Hummer, 1997) and the exploitation of the coherent radiation opportunities available at third-generation synchrotron sources.

12

Macromolecular crystals and twinning

Macromolecular crystal-structure analyses can be severely hampered by cases of twinning or of multiple-crystals. However, it is increasingly the case that twinning can readily be recognized notably when the X-ray diffraction data are analysed in detail, especially using intensity statistics. A very recent review of this topic is given by Helliwell (2008) covering not only twinning but also cases of other multiple crystals and lattice disorders, as well as possible remediation of such cases that is possible via alteration of crystal-growth conditions including use, mainly, of chemical additives. The case of multiple-crystal growths likewise can hamper crystal-structure analysis and although not necessarily associated with twinning is a crystal-growth situation where similar remediation methods can be adopted, or alternatively portions of crystals 'cut out' or tiny portions of crystals selected through use of X-ray 'microbeams'. There is also a nice body of case studies now in the crystallographic literature that are also covered in detail by Helliwell (2008). Superlattice effects in protein crystals are described in Chapter 13.

Twinning is the oriented association of two or more crystals of the same phase, which are related by an operation in direct space mapping the orientation of a crystal onto that of the other crystal(s) of the twin (Friedel, (1926); for a modern review see Hahn and Klapper, (2003)). Twins can form during crystal growth, following a phase transition, or as the result of a mechanical action. Here, we are interested in growth twins, which can be considered as a crystal-growth anomaly in the sense that the most stable situation, corresponding to a single orientation, is not achieved but instead crystals with different orientations coexist in the sample. When the different orientations are visible in the morphology, one commonly speaks of 'individuals'. More generally, each orientation is realized in a domain, but different domains (physically separated in a twin) may have the same orientation. The domains of a twin share, exactly or approximately, a common lattice, the *twin lattice*, that is in general a sublattice of the lattice of each domain: they are called *triperiodic twins* because a periodicity common to the domains forming the twin exists in the three dimensions of the crystal space. Cases in which this common periodicity exists in a subspace of the crystal space are called *diperiodic twins* and *monoperiodic twins* (Friedel, 1933): in the following they are referred to as *subperiodic twins*. When no common periodicity exists at all, which means that the orientation of the domains is not crystallographically related, one speaks of *bicrystals* (Hahn *et al.*, 1999).

Twinning can involve an imperfect macromolecule crystal-growth situation, including two or more crystals clustered together: the twin operation relates only approximately the orientations of the domains. The angular divergence is measured by the *obliquity* (defined originally by Friedel, (1926)) and twinning is by pseudo-merohedry, or by reticular pseudo-merohedry, when the lattice common to the domains is a sublattice of the lattice of each domain. Twinning often involves a perfect crystal-growth situation where the orientations of the domains are exactly related by the twin operation: twinning is then known as merohedry, or reticular merohedry. Cases of *metric specialization* are known too, characterized by coincidence of unit-cell parameters (e.g. $a = b$ in an orthorhombic crystal or $\beta = 90^\circ$ in a monoclinic crystal) and are known as *metric merohedry* (Nespolo and Ferraris, 2000). In any case, the twin operation is always a crystallographic operation about a direct lattice element (Law of Mallard: (Friedel, 1926)); the obliquity accounts for any deviation from exact mapping of the domains.

The typical domain sizes within a single twinned crystal are assumed to be small enough to be undetectable by visual examination (under an optical microscope) but large enough compared with the X-ray beam coherence length that scattered waves from separate domains do not create unusual interference effects. Whilst the vast majority of crystal habits have convex surfaces rather than re-entrant surfaces, the crystal habit showing such a re-entrant shape is a distinctive sign of twinning. Diffraction data reveals the twinning effects, if present, in detail.

12.1 Historical Examples

Wyckoff (1948) showed an electron micrograph of a 'highly twinned virus crystal' remarking that "In spite of the twinning shown by these crystals, a study of the molecular distribution over their side faces provides a good idea of their molecular arrangement." King (1965) described the crystal morphology of ribonuclease I crystals in detail. "Most of the crystals of ribonuclease I in any sample tube are usually intergrown (Figure 12(a)), only a minority being single (Figure 12(b)) and thus usable in X-ray diffraction studies. The intergrowths are usually composed of a series of individual crystals of ribonucleaseI related by rotation about the *c*-axis by about $6^\circ\ 20'$). While no simple twinning law was found to account for these intergrowths, their occurrence and that of the high-index faces may be related phenomena."

More recently Lott *et al.* (2003) remark, "Although cases have recently been reported of successful structure determination by MAD (Rudolph *et al.*, 2003; Yang *et al.*, 2000) or MIR (e.g. Terwisscha van Scheltinga *et al.*, 2003) using highly twinned data, in general the anomalous or isomorphous signal is sufficiently degraded by the errors introduced by detwinning so as to make structure solution impossible. This degradation increases as the twin fraction approaches 0.5, at which

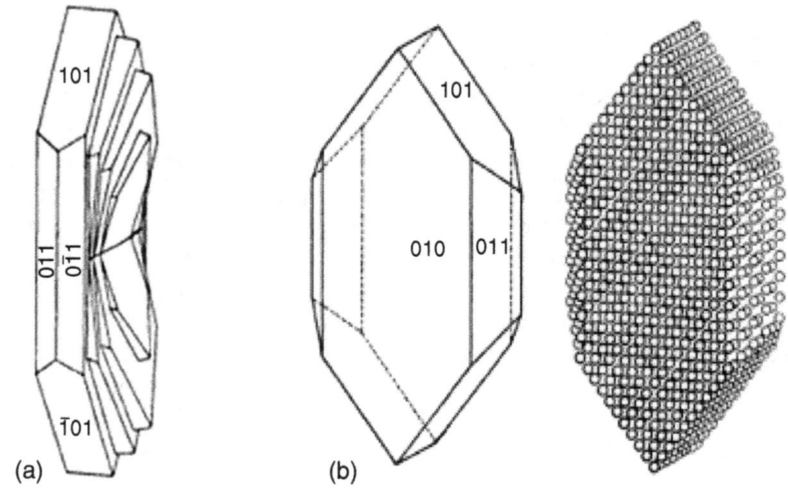

Fig. 12.1. Ribonuclease I crystals usually grow as (a) intergrowths but sometimes (b) as a single crystal (King, 1965); the circles (right figure) indicate the centroids of molecules without implication as to their shapes. Used with the permission of the IUCr.

point the true diffraction intensities cannot be deconvoluted from the measured twinned intensities." Likewise, Gulyaeva-Tcherkassova *et al* (2003) note that "Crystals of the Jel44 Fab fragment complexed with HPr diffracted X-rays to a maximum resolution of 2.5 Å . . . they were not characterized further owing to suspected crystal twinning." A further example is Donaldini *et al* (2003) who note that "The majority of crystals were twinned and it was difficult to identify single crystals solely based on appearance. Typically, only one crystal in ten was suitable for data collection, which severely hampered the search for heavy-atom derivatives. All attempts to reduce the level of twinning such as the use of additives (e.g. Hampton Research Additive Screens) were unsuccessful". Today twinning is regarded as a serious complication in structure determination and can even lead to a project being postponed until untwinned crystals are found.

12.2 Types of Twinning

Yeates (1997) has reviewed twinning in protein crystals: "In order to know when one is at risk, it is necessary to understand the symptoms of the different types of twinning and the symmetries where they may arise. There are two fundamentally different categories of twinning, epitaxial and merohedral. In epitaxial twinning, different domains are typically oriented so that their molecular spacings match at the crystal face where they meet, but their crystal lattices are not superimposable. In other words, their crystal lattices superimpose in fewer than three dimensions. This leads to distinct, interpenetrating reciprocal lattices in the diffraction pattern. Since epitaxial twinning is recognizable in the diffraction pattern, attempts can be made to integrate reflections from a single lattice, or untwinned crystals can be sought." The term 'epitaxial twinning' has been criticized by Nespolo (2009); who remarked that "these are nothing else than Friedel's subperiodic twins. Furthermore, 'merohedral' is an adjective that applies to a single crystal to indicate it lacks part of the symmetry of its lattice: to use it to describe twins may give rise to confusion. 'Merohedral twinning' refers in fact to the special cases where the lattices of the different domains overlap exactly in three dimensions: this is called *twinning by merohedry* because it can occur only when the crystal (not the twin) is merohedral; the adjective *merohedric* has also been introduced (Catti and Ferraris, 1976), precisely to avoid all confusions." Possible merohedric twin operators for tetragonal, trigonal, hexagonal and cubic space-group types are given as 'look-up tables' in Chandra *et al.,* (1999).

Merohedric twinning refers to two or more distinct lattices. Where twinning occurs, the more common case is hemihedral twinning where only two lattices are present. In this case, each observed diffraction intensity is then a sum of two crystallographically distinct, twin-related, diffraction intensities weighted according to their respective crystal volumes, assumed to be equally bathed in the X-ray beam.[1] The observed intensity is given by eqn (12.1) (Catti and Ferraris, 1976).

$$I_{\text{obs}} = \alpha I(hkl) + (1 - \alpha)I(h'k'l') \tag{12.1}$$

where hkl and $h'k'l'$ are the twin related indices and α is the 'twin fraction'. If the twin fraction, α, is equal or nearly equal to one-half, the observed diffraction pattern acquires an additional symmetry, i.e. as imposed by the twinning operation (see Figure 12.2 from Parsons (2003)). The diffraction data thereby possesses an erroneously high symmetry. This has been called 'perfect twinning'. The true crystallographic intensities cannot be recovered from these measurements, but in practice it has been found that it is still possible to solve a macromolecule structure, e.g. by molecular replacement, not least because the search molecule contains in effect a great deal of information independent of the diffraction measurements. The situation in which the hemihedral twin fraction is not one-half has been called 'partial twinning'. The result is not higher apparent symmetry, but observed intensities that still contain contributions from distinct reflections.

True crystallographic intensities can be extracted from a partially twinned crystal if the twin-fraction can be estimated accurately. This twin fraction, 'α', is estimated by a statistical evaluation of the similarity between twin-related observations. Pseudo- or non-crystallographic symmetry can be a complicating factor, as it tends in special cases to mimic twinning (at least at lower resolution) by leading to similarity between

[1] Of course for neutron protein crystallography involving a twinned crystal eqn (12.1) and the diffraction statistics equations quoted later, would apply just as they do with X-ray data. It is assumed here that a twinned crystal form for a given macromolecule would not be studied with neutrons; this may be a false assumption and tests for the occurrence of twinned diffraction data can of course be routinely applied as a precautionary measure.

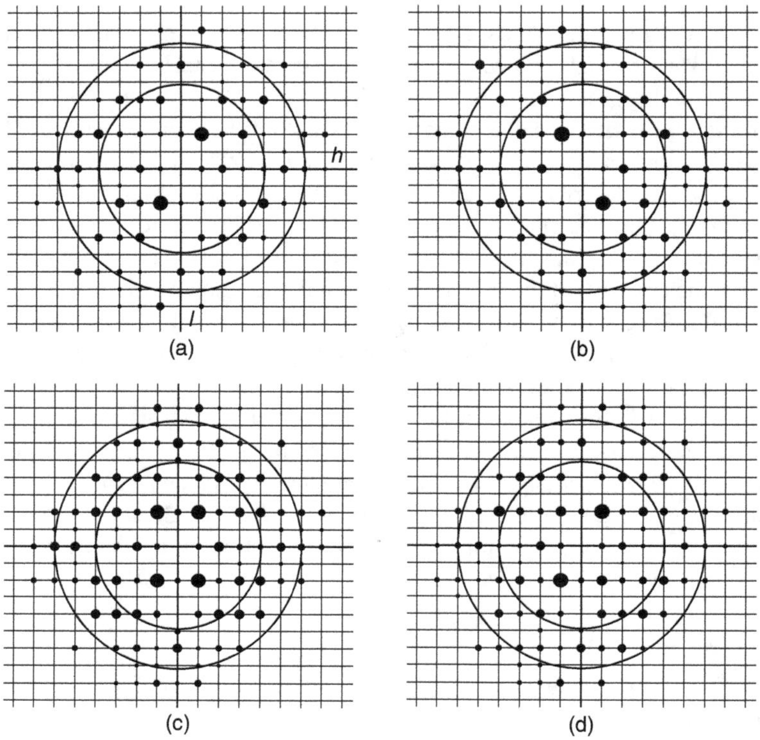

Fig. 12.2. A simulation of twinning and its consequences on the diffraction pattern (from Parsons (2003) with the permission of the author and of the IUCr). These show the effect of twinning by a two-fold rotation about **a** on the diffraction pattern of a monoclinic crystal with $\beta = 90°$, i.e. a case of twinning by metric merohedry. Only the h0l zone is shown; the space-group type is $P2_1/c$. (a) The diffraction from one domain only; (b) The diffraction from another domain (c) Superposition of the two domains in a twin with each domain of equal volume; (d) With the volumes reset to 80% and 20% of each domain in the twin. [The original article carries a more detailed caption to which the reader is referred if desired.]

crystallographically distinct reflections. Measurement errors lead to an incorrect estimation of the twin fraction, in practice, an overestimate, and so the test should be performed over a resolution range where the data are well measured.

Clues that diffraction data may be from a perfectly twinned specimen include: (a) a unit cell that is too small to contain the known molecule under the apparent space-group symmetry and (b) an intensity distribution that does not follow Wilson (1942) statistics. The latter as a test must be performed on normalized diffraction intensity data or in thin diffraction resolution shells. Anisotropic diffraction resolution can be a complicating factor. Examples of macromolecular software suites providing such statistics and their determination of twinning are discussed later.

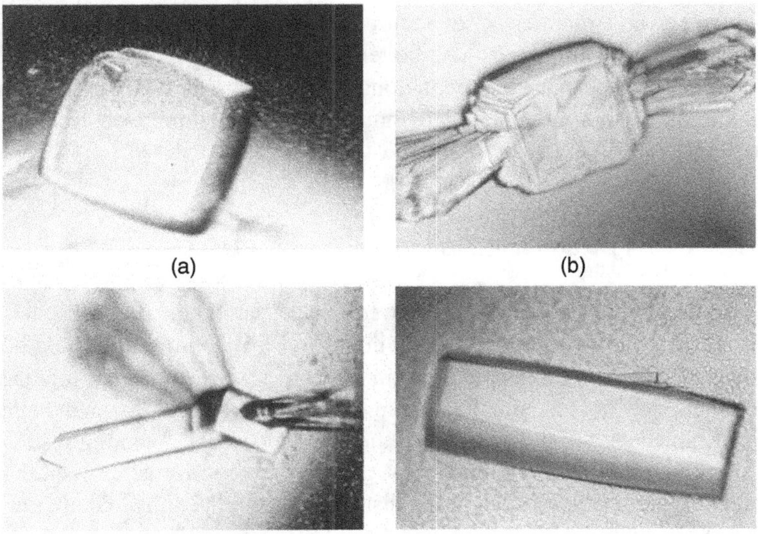

(a) (b) (c) (d)

Fig. 12.3. 'BJ-KWR' protein crystals grown at 295 K in (a) 1.5 M ammonium sulphate at pH8.0; (b) 1.45 M ammonium sulphate at pH 8.5; (c) 1.45 M ammonium sulphate pH 8.0; (d) 2.88 M NaH_2PO_4 and 0.7 M K_2HPO_4 pH 4.0. From Makino *et al.*, (2005) with permission of the authors and the IUCr.

Twinning can be detected at an earlier stage through examination of the crystals in an optical microscope. Good examples of this are given by Makino *et al.,* (2005) in their study of Bence–Jones protein and where "initially, crystals of a consistent though suspicious habit were observed. The suspect crystals, similar to those seen in Fig. 12.3(a), were "roughly cube-shaped and often grew to 0.2 mm or more in edge dimension where faces tended to be curved and the edges rather indistinct in some cases . . . they were physically robust and easy to mount for analysis. A second crystal habit formed and coexisted for some time with the first and then grew, apparently at the expense of the initial form. Frequently, this secondary crystal form nucleated on and grew from the surfaces of the cube-shaped crystals (Figs 12.3(b) and 12.3(c)), suggesting a possible unit-cell relationship. Occasionally, the prismatic form appeared directly without accompaniment by the first form."

Numerous attempts were made to collect diffraction data from crystals of the cubic habit with rounded faces but were unsuccessful. Often, some zones clearly showed the presence of multiple misaligned reciprocal lattices (Makino *et al.*, 2005). They further remark: "Nonetheless, it was sometimes possible to index these patterns and the unit cell was found to belong to space group $P4_122$ (or $P4_322$), with unit-cell parameters $a = b = 64.2$ Å, $c = 179.7$ Å. These crystals were the subject of a detailed atomic force microscopy (AFM) study (Plomp *et al.*, 2003), which demonstrated among other things that the unique axis was 4_3." Subsequently, Makino *et al.*, (2007) reported the solved structures of several of the crystal forms.

The presence of twinning can also be revealed by observation under a polarizing microscope, provided that the orientations of the domains do not give the same extinctions, *i.e.* in non-merohedric twinning. For an example see Kerfeld *et al.*, (1997).

12.3 The Twin Fraction and Testing for Twinned Data

Testing for twinning has been well reviewed (Yeates and Fam, 1999). Dauter (2003) provide an overview of the effect of twinning on diffraction intensities, how this twinning can be detected, and how the twin fraction can be estimated. From the initial indexing there may not be enough space in the unit cell to accommodate the required number of molecules. In the analysis of the diffraction data, twinning results in the averaging of unrelated pairs of reflections, and thus there is a smaller fraction of very weak and very strong reflections in the distribution of reflections than that normally observed. This can be used as an effective test for the presence of twinning (Stanley, 1972). If twinning can be detected, the twin fraction estimated and twin fraction is not equal to 1/2 then the data can be used to extract structural information. An early estimation for the twinning fraction was that of Britton (1972). Britton was analysing a crystal of diiodoacetylene, and made the general observation that "in twinned crystals the intensity contribution from the separate individuals can be estimated from the observed intensities without any knowledge of the actual structure, even when the two reciprocal lattices are completely superimposed". Britton analysed their data as a function of different twin fractions using the minimum R value as a target to determine the fraction. Fisher and Sweet (1980) adapted the Britton technique by plotting the number of negative intensity estimates as a function of the assumed twin fraction. A sharp break in the resulting curve, termed the Britton plot, indicates the upper bound of the twin fraction value. This works well with accurate recording of weak intensities but is less sensitive if that is not the case. Murry-Rust (1973) built on this noting that plotting twinned-related structure factors gives two bounding lines related to the twin fraction.

Another method for identifying twinning is the Rees $N(z)$ plot where z is the fraction of local average intensity and N is the number of reflections with intensities below this level (Rees, 1980). This was developed as a general theory of X-ray intensity statistics for twins by merohedry based on the mathematical analogy of the problem to a one-dimensional random walk (Rees, 1982). Probability distribution functions were derived (a) for intensities from a single twinned specimen, and (b) for the differences in intensities from two separate twinned specimens. Theoretical values of a discrepancy index between observed intensities from a twin, and calculated intensities from both correct and incorrect structural models were evaluated. A new method for determining volume fractions of crystals in a twin was also proposed, based on the distribution of differences in intensities between twin-related reflections.

Yeates (1997) introduced a twinning test such that for untwinned intensity data $\langle F\rangle^2/\langle F^2\rangle$, known as the Wilson ratio test, is 0.885, for a perfect twin and 0.785 for untwinned acentric data. Subsequently, Padilla and Yeates (2003) discussed the diagnosis of hemihedral twinning further, remarking that the statistics of local intensity differences are insensitive to the effects of anisotropic scattering and to the effects of pseudo-centring, effects that confound traditional intensity statistics. They introduced a parameter L,

$$L = (l_1 - l_2)/(l_1 - l_2) \tag{12.2}$$

which they show is more robust for discriminating between acentric and centric data and for detecting perfect twinning. They remarked that "because this new statistic appears to perform as well or better than traditional intensity statistics, the distribution of L should supplement or perhaps supplant some of the intensity analyses in previous use." The robustness of their approach was demonstrated with examples using real proteins whose diffraction data appeared aberrant by other methods of intensity analysis. The new statistic was therefore shown to be better suited than other available methods for diagnosing perfect hemihedral twinning.

Lunin *et al.*, (2007) have developed an advanced statistical model, which is designed to estimate the twinning fraction in merohedrally (or pseudo-merohedrally) twinned crystals. Their model takes experimental errors of the measured intensities into account and is adapted to the accuracy of a particular X-ray experiment through the standard deviations of the reflection intensities. The theoretical probability distributions for the improved model were calculated using a Monte Carlo-type simulation procedure. The use of different statistical criteria (including likelihood) to estimate the optimal twinning-fraction value was discussed. Their improved model enables better agreement of theoretical and observed cumulative distribution functions to be obtained and produces twinning-fraction estimates that are closer to the refined values in comparison to the conventional model, which disregards experimental errors. The results of the two approaches converged when applied to selected subsets of measured intensities of high accuracy.

Many of these approaches have been incorporated into software packages. Testing for the presence of twinning through detailed scrutiny of the X-ray diffraction intensity statistics in individual diffraction data sets can be readily made (Padilla and Yeates, 2003; Yeates, 1997). One of the most popular X-ray protein and macromolecule crystal-structure analysis packages is the CCP4 package (1994). Within this an early check/indication for twinning is the analysis of an X-ray data set intensity statistics in the computer program TRUNCATE.

The software 'PHENIX' (Adams *et al.*, 2004; Adams *et al.*, 2002) incorporates a component called xtriage (Zwart *et al.*, 2005) that performs a number of data-quality checks, which include a cell-contents analysis and a number of analyses related to twinning; twin operations are determined, the observed intensity distribution is compared to the expected distribution and an L-test is performed (Padilla and Yeates, 2003) as well as a Britton analysis and other tests described above. The analysis is presented in a summary that indicates the potential of twinning being present. The SHELX-97 software also processes twinned data (see Schneider *et al.,* (2000) for example): for details see Herbst-Irmer and Sheldrick (2002). In the most recent release of CCP4 (Version 6.1) the REFMAC protein model refinement program can now estimate twin fractions and helps where a protein model refinement has got 'stuck', e.g. with an R factor of say 35%.

Buts *et al.,* (2004) address the not uncommon situation that the only available crystals are aggregates of single crystals with different orientations, giving rise to two or more distinct but partially overlapping diffraction patterns. This situation is distinct from merohedric twinning, in which an exact symmetry relates the constituent lattices and the diffraction patterns overlap completely. Often, there are many different orientations contributing with similar intensities, yielding a very complex pattern that cannot be interpreted. In certain situations, however, it is possible to separately index two or more dominant patterns, resulting in one data set for each sub-crystal orientation. Since 'standard' data-reduction software is not aware of the presence of the other patterns' spots, which overlap with spots from another (crystal fragment) orientation, it will integrate to incorrect values for the spot intensities.

Lietzke *et al.,* (1996) developed methods to process diffraction data from super-periodic twins, whereby four programs for data reduction and two display programs were developed to augment the data-reduction program *XDS* (Kabsch, 1998). The programs, which the authors state can be generalized for use with other data-reduction software, provide the user with a list of the reflections used to determine lattice constants and crystal orientation. One such, *LATTICE_VIEW,* generates a file containing the reciprocal-space coordinates of the strong spots found in the initial data frames. The file is visualized to identify spots that belong to the same lattice, obtains unit-cell dimensions for a lattice, and assesses data quality. Their program, *VECTOR_MATCH,* is used to find additional spots belonging to a lattice. *ACCOUNT4* determines which spots have been processed by *XDS. COMFORT* discards reflections that are too close to a reflection in another lattice. The display programs provide useful visual information on the quality of the crystal orientations used.

12.4 Using Twinned Data

Case study; Detection and characterization of merohedric twinning in crystals of oxalyl-coenzyme A decarboxylase (Berthold *et al.*, 2006; Berthold *et al.*, 2007): Oxalyl-coenzyme A decarboxylase is a thiamin-diphosphate-dependent enzyme active in the catabolism of the highly toxic compound oxalate. Two crystal forms were obtained, one showing poor diffraction and the other merohedric twinning. The degree of twinning could not be reduced by chemical additives. Crystals in the former category belonged to a tetragonal space group of type $P4_22_12$. Data to 4.1 Å resolution were collected from these crystals and an incomplete low-resolution structure was initially determined by molecular replacement. Crystals in the twinning category yielded diffraction data to 1.73 Å collected from one of these crystals; the data had an apparent point group 622 but in fact was space group *P*321 and the crystal used was determined to have a twin ratio of 0.43. With the diffraction data treated in the lower symmetry and detwinned, a molecular replacement solution was obtained using the low-resolution structure as the protein molecule template. The structure revealed there was a non-crystallographic symmetry axis almost parallel to the twin axis, an arrangement that can promote twinning. Subsequently, Berthold *et al.*, (2007) were able to report refined structures of the enzyme in complex with different ligands and whereby crystallization with the enzyme in a ligand complex prevented the twinning effect. The new crystallizations were performed by the hanging drop vapour-diffusion method under conditions otherwise very similar to those described previously (Berthold *et al.*, 2006).

12.5 Overcoming Twinning in Crystals

Several chemical and molecular-biology approaches have been used to treating twinned crystals from simply changing the chemical conditions to a recent observation that affinity tags help with the problem in the case of membrane proteins (Backmark *et al.*, 2008). Microseeding is also now routinely used, e.g. Figure 12.4 and discussed in 'semi-automatic', i.e. pipeline project applications (Walter *et al.*, 2008).

Lagartera *et al.*, (2005) describe a 'Pce' esterase enzyme, whereby the first crystals of native Pce showed serious twinning problems. Their careful analysis of protein

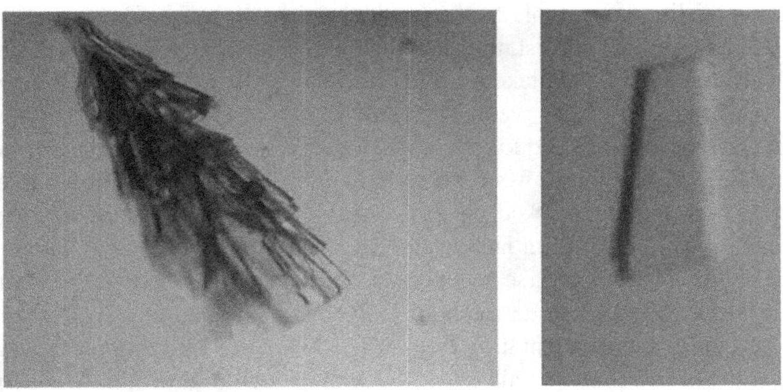

(a) (b)

Fig. 12.4. Single crystals of endoglucanase: (a) before optimization and (b) after optimization of crystallization conditions and using microseeding: the size of the single crystal in (b) is ~0.2 × 0.1 × 0.05 mm. Reproduced from Hirvonen and Papageorgiou (2002) with permission of the authors and the IUCr.

stocks by SDS ± PAGE and mass spectrometry showed a time-dependent appearance of two protein species corresponding to truncated forms of Pce (66.9 and 63.5 kDa) degraded at their C-terminal extreme. Analysis of Pce crystals by SDS ± PAGE using silver staining showed that only the 63.5-kDa species, which lacked the last 55 amino acids, crystallized. In order to improve crystallization, the last 55 amino acids were deleted, activity was maintained, and untwinned data resulted.

Sica *et al.,* (1994) described the use of agarose gels to slow down the process of crystallization and thereby achieved a positive impact on eliminating twinning. The method usually provides for the slow diffusion of the solute or solvent into the gel, and inhibits secondary nucleation, thus lowering convective flows and sedimentation effects. As their example, they showed crystals of an alcohol dehydrogenase 'SsADH (NADH)2', grown in agarose gels, and that lacked the clear physical twinning often observed in holoenzyme crystals grown by conventional means. This was confirmed by X-ray diffraction of a large prismatic crystal ($0.8 \times 0.4 \times 0.2$ mm), which gave a unique reciprocal lattice. Although growth in agarose gels eliminated the twinning phenomenon, unfortunately several of the crystals grown in this way completely failed to show any diffraction pattern whatsoever. Only crystals freshly extracted from the gelled droplet, and soaked in a stabilizing solution prior to excision from the gel and mounting, showed diffraction. Another method of reducing convective flows has been the use of the microgravity environment, recently reviewed by Snell and Helliwell (2005). There are too few case studies on the effect of this environment on twinning to draw systematic conclusions but several cases are reviewed by Helliwell (2008).

A common practical procedure is cutting out a crystal either by mechanical force or more elegantly, by biochemical means. Efremov *et al.,* (2004) describe the splitting of hexagonal hemihedrally twinned bacteriorhodopsin (bR) crystals by a slow reduction of the salt concentration in the mother liquor. The split parts displayed high diffraction quality comparable to the original twinned crystal. They describe their procedure as follows, "At first, the attempts to separate agglutinated plates applying mechanical force were undertaken. However, the crystals cracked across the *ab*-plane before the hexagonal plates were separated. Therefore another approach based on gradual changes of mother liquor properties was used. It was observed that whenever a bR crystal was transferred from 3M sodium phosphate buffer to the solution with low salt concentration it cracked across the hexagonal plane into small segments and at the same time the hexagon laminated. This observation suggested that a gradual change of salt concentration can induce a crack along the surface between twin domains before the domain destruction. Indeed, it was found that a slow reduction of mother liquor salt concentration from 3 to 1 M induced splitting of agglutinated plates. Some of the split crystals diffracted well enough to determine the twin ratio which in all cases was equal to zero within experimental error." The nature of the twinning at the molecular level was examined, to quote "The crystal twinning domains are in contact with each other whereby there are two possibilities for the adjacent proteins in the crystal layout to contact each other: by cytoplasmic to cytoplasmic (CPtoCP) or by extracellular to extracellular (ECtoEC) surfaces of bR.

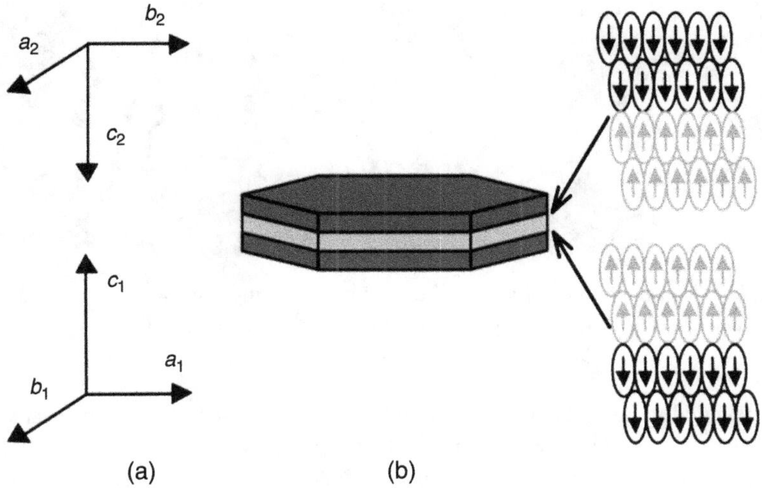

Fig. 12.5. The case of bacteriorhodopsin (BR). (a) The mutual orientation of the crystallographic axes in a hexagonal crystal of bR and (b) The domain organization and schematic packing of bR molecules on the surfaces of the crystal sample's twinning domains at their contact. From Efremov *et al*., (2004) with permission of the authors and the *Biophys J.*

The organization of the crystals may originate from the difference in attraction of twin domains for the contact of the ECtoEC and CPtoCP molecular surfaces. The difference in this interaction may arise due to the difference of charge distribution on the water-exposed surfaces of bR. The extracellular surface of bR is neutral, whereas the overall charge of the cytoplasmic surface is negative. The interaction of short loops inside a crystal provides a relatively weak van der Waals contact between the layers: (the crystallographic structure reveals only van der Waals contacts between two amino acid residues per molecule). Hence, even a weak electrostatic repulsion of cytoplasmic 2D crystalline surfaces may have a noticeable impact on the interdomain interaction."

MacRae and Doudna (2007), worked on the crystal structure of the enzyme 'Dicer' from *Giardia intestinalis* to 3.3 Å resolution by MAD using crystals belonging to space group $P2_12_12$ (MacRae *et al*., 2006). These crystals were derived from crystals that initially diffracted X-rays to 3.0 Å resolution, but were described as "refractory to structure determination because they were twinned" Manual dissection of the crystals was described by the authors as "difficult and the crystals were heavily damaged during the process". The partially dissected crystals still diffracted X-rays sufficient for crystal-system assignment. The authors collected diffraction data on one lobe or the other of the sample, and demonstrated that data measured from either individual lobe had orthorhombic symmetry, whereas data measured from the lobe junction, which contained overlapping contributions from both lobes, showed tetragonal

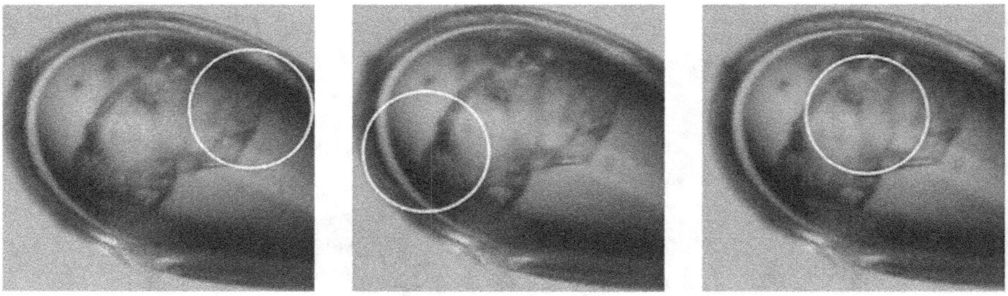

Fig. 12.6. Selecting portions of a 'Dicer' crystal sample using a small SR X-ray beam. From MacRae and Doudna (2007) with permission of the authors and the IUCr.

symmetry analogous to the data measured from the original crystals. By transferring Mn-grown crystals into less concentrated precipitant solutions, the macroscopically twinned crystals were partially dissolved. This treatment, similar to that used to dissect hemihedrally twinned crystals of bacteriorhodopsin (Efremov *et al.*, 2004) weakened the connection between each lobe, allowing the crystals to be readily dissected into their component single crystals for diffraction data collection.

The original Dicer crystals represented an unusual case of perfect pseudo-merohedral twinning of orthorhombic crystals. Before the twinning problem was overcome, it was possible to calculate a low resolution electron-density map in space group $P4_1$ that was used to build a partial molecular model. Experimental phases were sufficient to identify heavy-atom sites that indicated space-group inconsistency, leading to the efforts to identify the true space group. This information guided the search for different crystallization conditions that yielded untwinned crystals and ultimately a fully interpretable electron-density map.

If none of the above remediation methods proves possible, then what to do? MIR and MAD or SAD phasing can be severely disadvantaged by twinning, working as they do with small structure-factor-amplitude differences for phasing. In particular, a researcher employing MIR on a given project, working with different native and heavy-atom derivative crystals, may have to face the fact that different crystals have different twinning fractions. Obviously, MAD and SAD often involve use of a single crystal and is an advantage over MIR. Indeed Yang *et al.*, (2000) report successful MAD protein structure determination for a twinning fraction of 0.36, even with the twinned data. Their anomalous difference Patterson sections showed clear seleno vectors, as well as some vectors for the minor twin fraction, and showed the correct Harker sections. Furthermore, their electron-density map was improved upon working with detwinned data. There can still be the headache for MAD experiments using different regions of the same crystal where the twin fraction might vary along the crystal or, more obviously, where more than one sample is required due to radiation damage sensitivity limiting the amount of data per crystal sample.

Perhaps more optimistically though is the increasing chance to use molecular replacement as protein fold space coverage steadily improves. This has the advantage that it is a more robust situation to start with a reasonable molecular model for structure solution with the X-ray data from the new, twinned, crystal. One detailed case study of use of molecular replacement, involved a twin fraction of 0.43 to solve a site-specific mutant of the N-terminal half-molecule of human lactoferrin (Breyer *et al.*, 1999). Obviously this is a case of very high amino acid sequence identity rather than the 'twilight zone' cases of much lower sequence identity in the 30% range and for which an instance of twinning would be a very serious complication to face as well as the low biochemical similarity of two molecules. In the case of this mutant of lactoferrin, Breyer *et al.*, (1999) conclude on an optimistic viewpoint, "Finally, even in the case of perfect hemihedral twinning, it should still be possible to solve complicated molecular-replacement problems. This can be performed by examining the behaviour of the correlation coefficient as the positioned fragments are systematically permuted among the twin domains".

An approach now becoming more practical is the use of a small X-ray beam to select one component of a twinned crystal. This sort of approach involves using an X-ray beam smaller than a crystal and invariably requires use of intense and well-collimated synchrotron radiation (SR). This proved a good application of 2nd-generation SR sources and with even smaller beams using undulator X-radiation at 3rd-generation SR sources. For an early example of use of a 2nd-generation source to demonstrate the recording of good signal-to-noise diffraction from a ~20 micrometre gramicidin crystal, see Hedman *et al.,* (1985) using SRS Station 9.6 (Helliwell *et al.*, 1986b) in 'Laue mode'.

ESRF and EMBL-Grenoble originated the now so-called 'microbeam diffractometer' (Perrakis *et al.*, 1999): i.e. for use with X-rays in micrometre-sized beams (typically 10 micrometre but now as low as 1 micrometre). For an example of use of a small, i.e. untwinned crystal portion on this experimental set up, see Renault *et al.,* (2001). Xiao *et al.* (2003) in their crystal-structure determination of the retinoblastoma tumor suppressor protein bound to E2F targeted a crystal portion using the ESRF 10 to 20 micrometres beam and thus found a much lower crystal mosaicity suitable for data collection.

12.6 How Prevalent is Twinning?

Entries deposited in the Protein Data Bank as of February 2004 for which both model and X-ray data were available were analysed by Lebedev *et al.,* (2006) to identify cases of twinning using such simple statistics as the *R* factor between potential twin-related reflections. Careful consideration of all identified twins showed that in quite a number of cases twinning was ignored during structure solution and refinement. Manual analysis of the models showed that twinning often occurred in association with rotational pseudo-symmetry parallel to the twinning operator. The co-existence of these two phenomena complicated the detection and diagnostics of twinning using currently available twinning tests. It was concluded that a twinning-detection step

Table 12.1. Frequency of twinning in different symmetry environments (i.e. types of space groups). From Lebedev *et al.*, (2006). With permission of the authors and the IUCr.

Crystal symmetry or type of twinning	Number of twins				
	Total	RPS	TNCS	RPS + TNCS	DNA
P1	1	-	-	-	1
$P2_1$	13	11	3	3	-
C2	1	-	-	-	-
$P2_12_12$	1	-	-	-	-
$P2_12_12_1$	2	2	-	-	-
$C222_1$	1	1	1	1	-
Pseudo-merohedric total	19	14	4	4	1
$P4_1$	4	1	1	-	-
$P4_3$	6	4	1	1	-
$P4_2$	1	1	1	1	-
I4	3	2	-	-	-
$P3_1$	7	5	-	-	-
$P3_2$	6	4	2	1	-
P321	2	2	-	-	-
$P3_121$	2	-	-	-	-
$P3_221$	1	1	-	-	-
$P3_212$	2	1	-	-	1
H3	15	3	1	1	2
$P6_3$	6	3	-	-	-
$P6_5$	3	3	2	2	-
$I2_13$	1	-	-	-	1
Merohedric total	59	30	8	6	4
Total	78	44	12	10	5

RPS = Rotational pseudo-symmetry
TNCS = Translational non-crystallographic symmetry
DNA = Deoxyribonucleic acid cases

should be incorporated in every stage of structure analysis from data acquisition to refinement and validation.

12.7 Chapter Summary

There is now a healthy literature on crystal twinning and it is well documented that crystal-structure analyses can be severely hampered by the condition. It is increasingly the case that twinning can be diagnosed. Twinning is a situation where the macromolecular crystallographer might well step back to crystallization or the purification of a protein or in more extreme cases to systematically altering the target protein. Moreover, methods for the remediation of twinning cases, notably via alteration of crystal-growth conditions including use of chemical additives, have steadily become known. Tailored apparatus and small X-ray beams have also been harnessed to work with the crystals that one has and new physical approaches still emerge, e.g. for preparing protein crystallites such as using 'jet sprays' or of fs lasers for precisely and delicately cutting up crystal clusters into individual pieces. As the throughput of macromolecular crystal structures continues to increase the likelihood of crystallographers facing twinning and/or other disorder cases also increases. Furthermore, as protein fold space coverage steadily improves the chance for harnessing an appropriate molecular replacement model also improves. Thus whilst MIR and MAD or SAD phasing can be severely disadvantaged by twinning, working as they do with small structure-factor-amplitude differences for phasing, the advantage of molecular replacement is that a more robust starting situation holds for structure solution with the X-ray data from a new, but twinned, crystal using an existing, precise, structure model.

13

Other macromolecular crystal diffraction disorders

13.1 Introduction

The literature on other macromolecular crystal diffraction disorders is much smaller than that involved with twinning. Lovelace *et al.,* (2008) most recently described an incommensurate superlattice (Figure 13.1). Pickersgill (1987) described an early example involving a crystalline ternary complex of spinach ribulose 1,5-bisphosphate carboxylase/oxygenase (rubisco) with activators Mg^{2+} and CO_2. This diffracted to high angle but showed diffraction features symptomatic of disorder. These features, intensity streaks, were along c^* and were resolved into satellite reflections, with underlying diffuse scattering, by employing synchrotron radiation on SRS station 9.6 (Helliwell *et al.*, 1986b) with slits especially set to limit the horizontal beam convergence from the monochromator thereby giving a fine collimation (Figure 13.2). The disorder was described by Pickersgill (1987) in terms of stacking faults between layers of molecules, with every layer randomly shifted by either $a/39$ or $-a/39$ with respect to the previous layer, giving intensity streaks for the maximally affected reflections. The observation of superlattice reflections and broadening was best accounted for by considering an approximate superlattice and by additional shifts of $\pm a/39$ leading to the diffuse features.

While reports of these types of disorder are rare there are a number of case studies that are useful to explore. This data should not be discarded but can provide useful information when treated correctly and in some cases may be misinterpreted as a badly diffracting crystal.

13.2 Case Studies

Case study 1; An archetypical order-disorder crystal-packing arrangement in biology Mancini *et al.,* (2003) studied the special case of crystal-packing arrangements and disorder of hexameric NTPases from dsRNA bacteriophages. They reported that “Crystals of full-length ‘P4_8’ belonged to the apparent space group $P6_322$, with unit-cell dimensions consistent with the presence of one monomer of the P4 protein in the crystallographic asymmetric unit.” However, a rather complex diffraction pattern was seen (Figure 13.3) that included a honeycomb-like effect

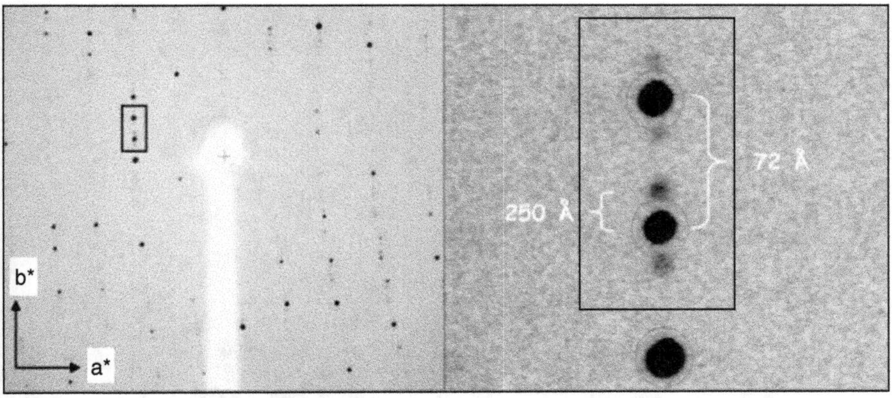

Fig. 13.1. A profilin:actin single crystal, showing an incommensurately modulated diffraction with the b*-direction lattice and superlattice unit-cell dimensions shown (Lovelace *et al.*, (2008) with permission of the authors and the IUCr).

and diffuse rods and bands in addition to the Bragg reflections. Mancini *et al.* noted that these effects were characteristic of statistically disordered crystals, first reported in 1954 for imidazole horse methaemoglobin (Bragg and Howells, 1954; Cochran and Howells, 1954).

In this example, crystal domains showing systematically different orientations or positions are distributed randomly throughout the crystal causing a characteristic systematic blurring of the Fourier transformation. The diffraction seen was explained with the crystals being based upon a perfect network of hexamers organized in a smaller unit cell, forming two-dimensional layers with well-defined structure that yielded sharp reflections. These were then stacked together with random fractional displacements of 1/2 in the *a*- or *b*-direction; manifesting as an apparent doubling of the unit-cell dimension in the *c*-direction. The reflections of this superlayer were diffuse, leading to the sharp and diffuse lines of various widths.

This phenomenon is known as 'OD', for 'order-disorder' and leads to polytypes, in which the stacking of layers may result in a definite periodicity (ordered polytypes) or, as in this case, to a lack of periodicity in the direction of stacking (disordered polytypes). A subset of reflections, common to all polytypes, are always sharp and are called 'family reflections'; the rest of the diffraction pattern is composed of the so-called 'non-family reflections' or 'characteristic reflections' and are affected by streaking when the polytype is disordered. The theory of OD structures was developed by Dornberger-Schiff and her coworkers (see, e.g., Dornberger-Schiff, (1979; Ďurovič, 2004)) and is widely applied to inorganic and mineral compounds, although OD phenomena exist as well in molecular and macromolecular crystals.

Case study 2; A weighted sum of two identical but translated structures, one from each lattice and the dangers of obtaining a biologically irrelevant complex Wang *et al.*, (2005) faced an example of lattice-translocation defects (Figure 13.4), i.e. whereby two identical but translated lattices coexisted as a single coherent mosaic

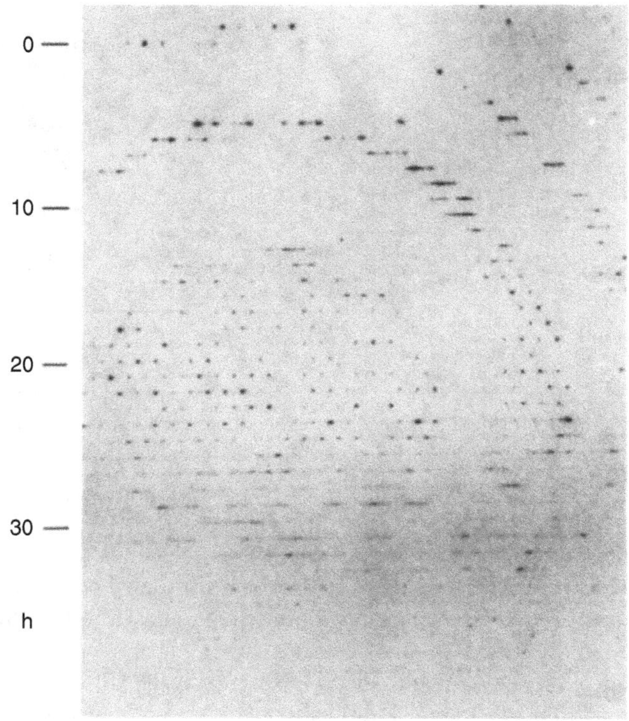

Fig. 13.2. Enlargement of part of an oscillation photograph from a Rubisco crystal taken around the c*-axis showing satellite reflections and diffuse scattering in the regions of h equal to 10 and 30. This photograph was taken using the Daresbury Synchrotron Radiation Source 'SRS9.6' ($\lambda = 0.88$ Å, 1.8 GeV, 235 mA). In order to resolve the satellite reflections a 0.2-mm collimator was used and the horizontal beam convergence was limited to 0.6 mrad. The crystal-to-film distance was 144 mm corresponding to a resolution at the edge of the film of 2.28 Å. An oscillation range of 1.5 ° and exposure time of 75 min were used (recorded by R.W. Pickersgill and J.R. Helliwell).

block in a crystal. The observed structure in such cases is a weighted sum of two identical but translated structures, one from each lattice: the observed structure factors are a weighted vector sum of the structure factors with identical unit amplitudes but shifted phases. The correction of X-ray intensities from a single crystal containing these defects of the 'hybrid HslV-HslU complex', which consists of *Escherichia coli* HslU and *Bacillus subtilis* HslV (also known as CodW), was reported. When diffraction intensities were not corrected, a biologically irrelevant complex (with CodW from one lattice and HslU from another) was obtained. Only upon correction did a biologically functional CodW-HslU complex structure emerge.

Case study 3; Multiple allowed packings of an AB_5 hexamer and where it was possible to ignore the superlattice Merritt, *et al.,* (1997) encountered a superlattice created by multiple allowed packings of an AB_5 hexamer. The primary lattice contacts

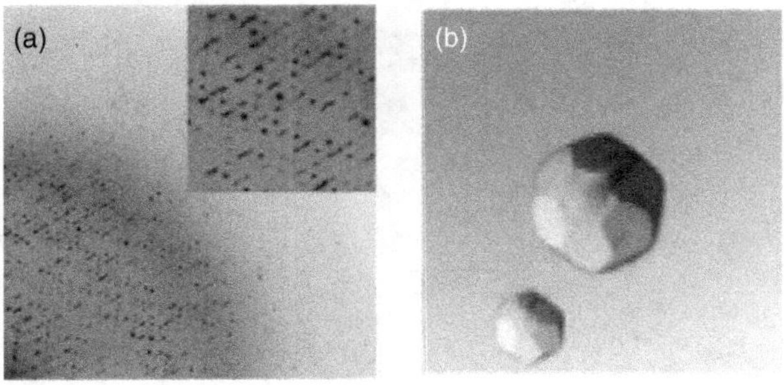

Fig. 13.3. The rather complex diffraction pattern (a) of a crystal (b) for the NTPase from dsRNA bacteriophage that included e.g. diffuse rods and bands (inset). From Mancini *et al* (2003) with permission of the authors and the IUCr.

were defined by the B_5 pentamer packing interactions. Considered in isolation, the A protein subunit could be oriented in any one of 5 settings about the pentamer's 5-fold. In context, the choice of an A orientation in one cell limited the compatible choices in the adjacent cells. This led to a preference for sequential orientations of A in adjacent cells, and hence a superlattice. Merritt (personal communication) also states that "The diffraction images could be indexed and integrated using various multiples of the minimal axis length along that one direction. The strongest correlation of adjacent cells had a repeat length of 3 along one axis, so the most evident superlattice in the diffraction images was an apparent tripling of that axis. Since the supersymmetry was stochastic, and involved a 5-fold rotational axis, you may imagine that the diffraction patterns had other interesting features as well. In the end it was decided to ignore the superlattice, and refine against only the minimal set of Bragg reflections. This produced maps in which the A subunit was multiply ordered about the 5-fold, with 3 predominant contributing orientations. But since the interest was in ligand binding by the B_5 pentamer this did not prevent interpretation of the relevant binding sites."

Case study 4; Structure solution example with or without correction to the measured intensities Zhu *et al.,* (2008) determined the structure of N1 neuraminidase derived from influenza virus. 'The crystal was indexed in space group $C222_1$, with unit-cell parameters $a = 117.7$, $b = 138.5$, $c = 117.9$ Å, and the structure was solved by molecular replacement. The lattice-translocation vector in the 18NA crystal was (0, 1/2, 1/2) or its equivalent vector (1/2, 0, 1/2) due to the C lattice symmetry. Because of this special lattice-translocation vector in space group $C222_1$, the structure refinement could be achieved in two different ways: using corrected or uncorrected diffraction data. A composite model was built to represent the molecules in the translated and untranslated layers, respectively. This composite structure model provided a unique example to examine how the molecules were arranged in the two lattice domains resulting from lattice-translocation defects.' The diffraction patterns in two different orientations are shown in Figure 13.5 together with the structures, nicely demonstrating how one can imagine how the crystal growth might occur at the molecular level.

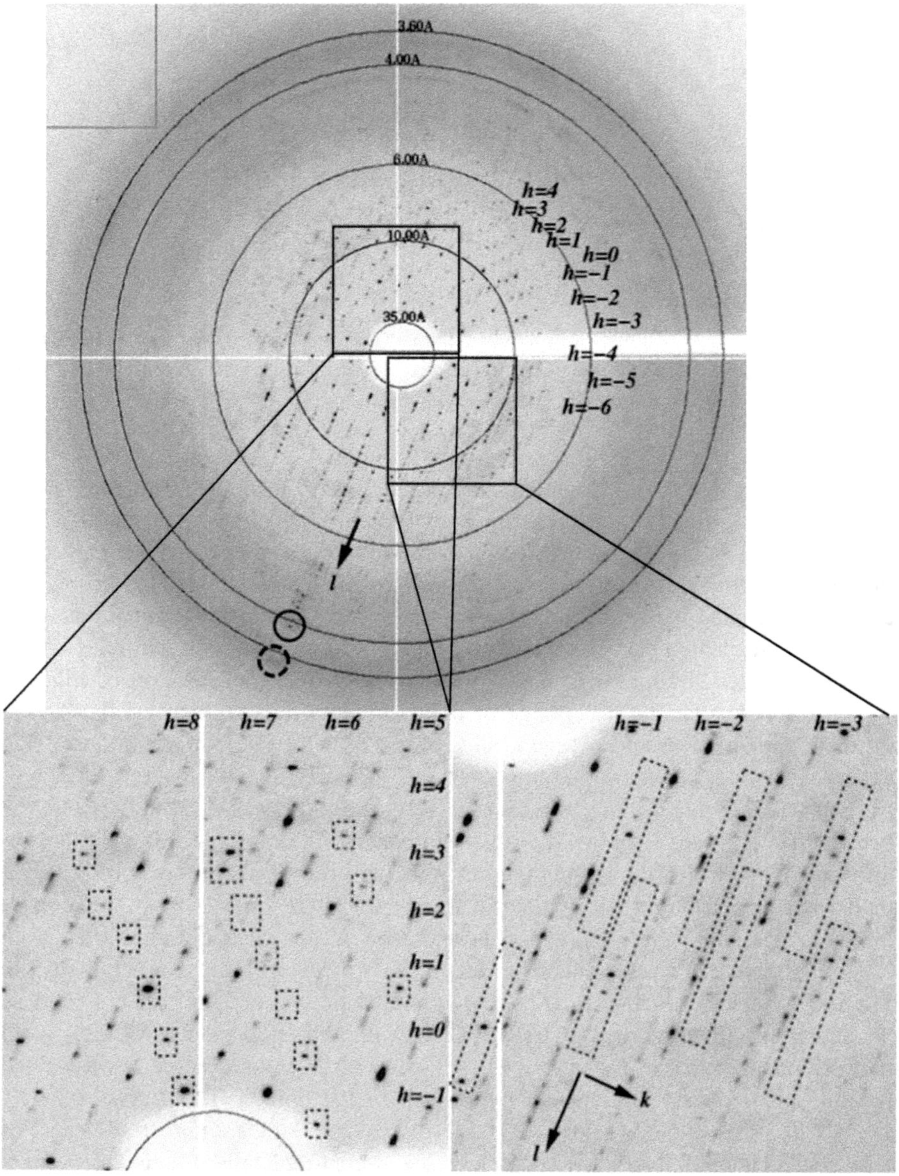

Fig. 13.4. Observed alternating sharp-diffuse reflections from a crystal of a 'hybrid HSIV-HSIU' protein complex. In the *l*-direction, strong reflections are observed beyond 4.0 Å and some reflections are observed at 3.6 Å; in the *h*- and *k*-directions reflections barely pass 4.5 Å. Only the third of the reflections with index $l = 3n$ (boxed) have typical Bragg spots and the remaining reflections have streaky features along the l-direction. Data were processed to 4.16 Å, slightly below an ice ring at 4.15 Å. From Wang *et al.*, (2005) with permission of the authors and the IUCr.

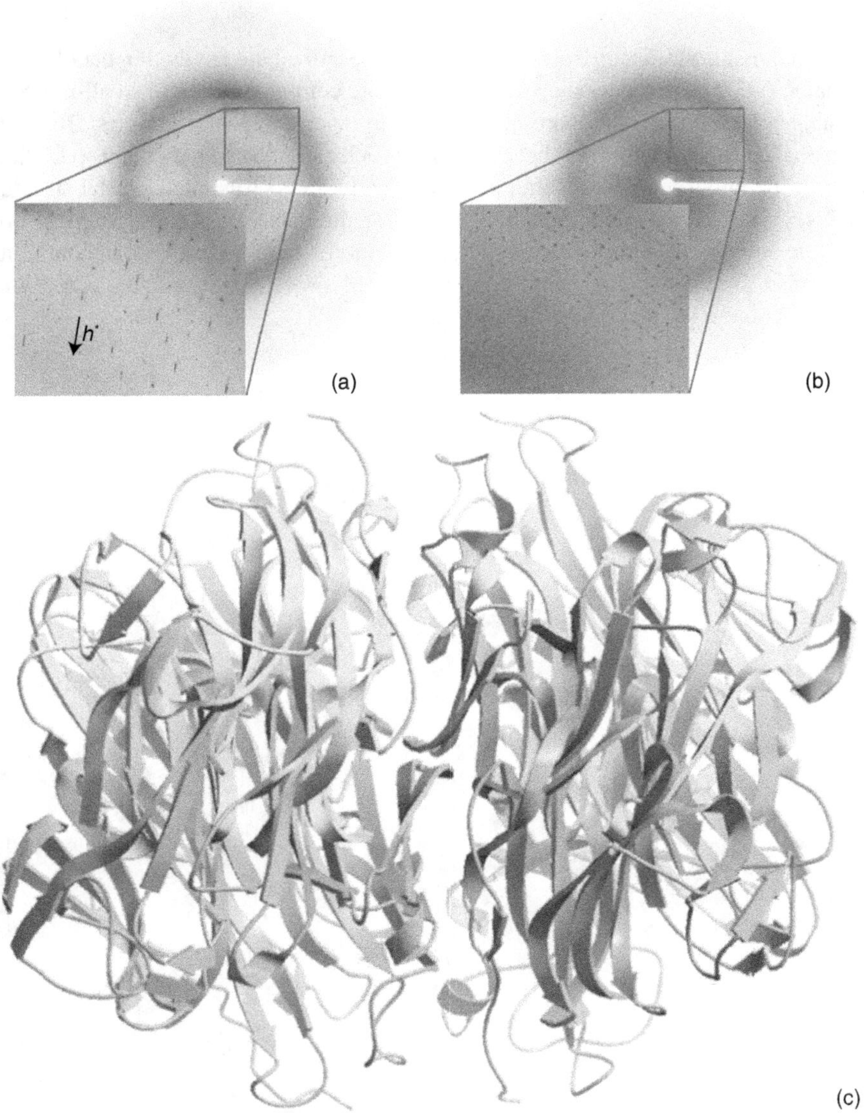

Fig. 13.5. Diffraction patterns and structure of a native 18NA crystal with lattice-translocation defects (from Zhu *et al.* (2008)). (a) Alternating sharp–diffuse reflections were observed and the streaky axis of the smeary features was determined to be parallel to the b*-direction, which is nearly vertical in the plane of the paper. (b) The b*-direction was found to be nearly parallel to the X-ray beam and perpendicular to the plane of the paper. The sharp–diffuse sets of reflections were not obvious in this orientation. (c) 18NA molecules in ribbon representation in the case when treated as two lattice domains and showing the high-occupancy and low-occupancy dimers (green and grey in the original publication, respectively) superimposed. With permission of the authors and the IUCr.

13.3 Chapter Summary

The fundamentals of disorders such as the ones described are covered in detail by van Smaalen (2007). Disorders such as these are more commonly seen in small-molecule crystals although the same theory is pertinent to macromolecular samples. Borgstahl (personal communication) has been gathering such examples together and so far there seem to be about ten cases, some published and others 'frustrated crystal-structure analyses', i.e. as yet uncompleted. The case studies above already document how possible solutions allow detailed structural data to be obtained and are encouraging. They also show that data sets initially deemed problematic and discarded may actually contain significant useful information.

14

Degradation and improvement of crystal perfection

The diffraction properties of a crystal can be degraded by a number of factors. These include age, X-ray radiation damage, cryo-cooling to mitigate this damage, and dehydration effects. Some, initially unexpected, manipulations of crystal perfection have improved diffraction quality and mosaicity such as through controlled dehydration and annealing. In this chapter we address these factors.

14.1 Ageing

Age can deleteriously affect a macromolecular sample in solution. For example, in the case of a membrane protein system, photosystem II, only freshly purified non-frozen protein ever produced crystals (Smatanova *et al.*, 2006). Similarly for a more prosaic sample, lysozyme, aged solutions display a tendency to produce more but smaller crystals than the fresh case (Chayen *et al.*, 1993). Clancy *et al.* (1994) report that crystals of an interlukin-1 receptor antagonist protein gradually lose the ability to diffract X-rays over a period of two months. In a more critical case for succinate:ubiquinone oxidoreductase, diffraction only occurred from crystals cryo-cooled within 72 h of the crystallization setup; any longer than that and the crystals did not diffract (Horsefield *et al.*, 2003). A common sense approach is to make use of crystals as soon as they reach a volume suitable for X-ray diffraction i.e. to collect data on them or to cryo-preserve them at this point (described below). While using the crystals rapidly seems good practice, in the extreme case one may want to wait until technology, techniques, and software catch up with the problem. Jovine *et al.*, (2000) report the redetermination of a yeast phenylalanine tRNA structure from 15-year-old crystals. They thereby resolved the bound-water structure in the Mg binding sites and determined the role of water in coordinating the metals to the tRNA. Similarly, Simonovic and Volz (2002) report the structure of a succinimide intermediate in *E. coli* from crystals that were 8 years old.

14.2 Radiation Damage

As early as 1976 Blundell and Johnson (1976) brought together the effects of radiation damage; they covered topics on comparison of radiation damage effects of electrons and X-rays as well as the general absence thereof with neutrons, being a neutral probe. Experiences with X-radiation damage chemical scavengers were also listed. Sample heating as another effect of absorption of X-rays is mentioned as a concern. Helliwell (1984) gives detailed adiabatic (in isolation) calculations of this sample heating effect as well as calculated sample X-irradiation lifetimes in the context of the arrival of 3rd-generation synchrotron X-ray sources of that time. The successes of these sources in macromolecular crystallography have been spectacular. So much so that provision of such beamlines now accounts for ~40% of the APS and ESRF insertion device sectors. Nevertheless X-radiation damage has been a continuing concern not least as the 'cutting edge' of capability to study ever smaller crystals and ever larger unit cells at synchrotron radiation sources has improved and been optimized. A notable development has been a series of International Radiation Damage Workshops on Macromolecular Crystal Radiation Damage Studies. There have been five of these workshops to date with collections of papers appearing in the *J Synchrotron Radiation*. Progress in the area, i.e. what is now known and questions that remain to be answered, is summarized in a report on the fifth workshop in Garman and Nave (2009). Four of the five workshops have resulted in special issues of the *J Synch Radiat* (Volumes, 9, 12, 14 and 16, parts 6, 3, 6 and 2, respectively).

Elastic scattering, which provides the diffraction spots pattern, is ~0.1% of the total incident X-rays at 8 keV for a typical (~0.3 mm on edge) macromolecular crystal. Inelastic scattering or absorption, i.e. Compton scattering and the photoelectric effect, is ~0.16% of the incident X-rays and forms the background 'noise'. Garman and Owen (2006) note that "1.7% of the incident X-rays are directly absorbed by the crystal". This depends critically on the size and shape of the crystal, as well as its orientation in the beam and its detailed chemical composition, and the X-ray wavelength used. The remaining X-rays do not interact with the sample. The absorbed X-rays cause ionization, forming electron-loss centres together with secondary electrons (O'Neill *et al.*, 2002). The primary radiation damage, formation of the electron-loss centre, is dependent only on the energy and number of photons absorbed; it cannot be prevented. The secondary electrons induce further excitation and ionization events and can create more free radicals. The migration of species created by the secondary electrons can be mitigated by cryo-cooling (covered in the following section) and by the introduction of free-radical scavengers (surveyed by Murray and Garman, 2002). Most chemical studies on the irradiation of macromolecules have focused on aqueous solutions at ambient temperatures (Houvee-Levin and Sicard-Roselli, 2001). Here, water radicals induce the majority of the damage. Low-temperature effects are not as well studied. The reactions of water radicals are varied. Oxidation of amino acid residues by hydroxyl radicals follows the order tryptophan > tyrosine > thiols > disulphides ~ aliphatic residues (O'Neill *et al.*, 2002). Even at cryogenic temperatures, amino acids may still be deprotonated, and decarboxylation may occur. Carbonyl groups and disulphide bridges may trap electrons resulting in subsequent fragmentation. Aromatic residues and metal redox centres are also sites of charge trapping where structural change may occur; a notable challenge is the Mn_4Ca oxygen evolving complex in photosystem II (for a review see Barber (2009)).

Southworth-Davies *et al.* (2007) have reviewed radiation damage at ambient temperatures extensively in a study on dose rate and radiation damage. We repeat some of their observations here. They observed that macromolecular crystals collected at room temperature can have longer dose lifetimes at synchrotrons than using a rotating-anode X-ray source (typical laboratory X-ray generating equipment); there is a dose dependence on radiation damage at room temperature. One should note that if data collection at ambient temperatures is attempted at the synchrotron, active temperature control should be used to avoid heating effects from the beam (Snell *et al.*, 2007 and references therein).

Blake and Philips (1962) carried out the first systematic study on radiation damage and proposed a model based on their data. They proposed that the irradiated crystal consisted of three populations; undamaged macromolecules responsible for the remaining high-resolution diffraction, a highly disordered fraction still contributing to low-order diffraction, and a destroyed part that contributes no coherent diffraction at all. The ratio of these fractions changes as a function of dose, Figure 14.1. Hendrickson (1976) derived an analytical expression for radiation damage based on all possible paths in the Blake and Philips model that explained experimental data

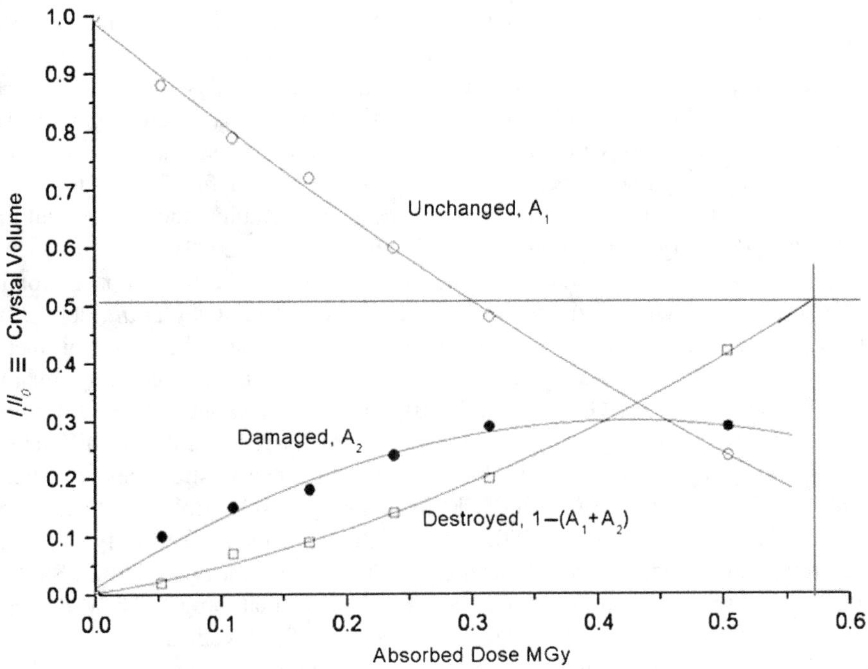

Fig. 14.1. Results of Blake and Phillips' measurements on crystals of myoglobin (from Southworth-Davies *et al.*, (2007) with permission of the authors and Structure.

well at moderate dose but broke down at higher doses. The results of Southworth-Davies *et al.* (2007) agree with Blake and Philips (1962) but show that "a longer crystal lifetime is obtained by using a higher dose rate, with a 60% increase in the dose rate leading to a 4-fold increase in the dose required for the summed diffraction intensity to fall to half of its original value". These results suggest that room-temperature data collection on a synchrotron source should not be overlooked, and indeed was routine in the past before cryo-crystallography became popular in the 1990s onwards.

Cryo-cooling (described in more detail in the next section) is routinely used to mitigate X-ray (and electron) radiation damage (Hope, 1988) but is not completely successful at the synchrotron (Gonzalez and Nave, 1994; Gonzalez *et al.*, 1992). With the advent of intense third-generation synchrotron sources, radiation damage even to cryo-cooled macromolecular crystals is seen as a significant and undesirable component of the experiment (Garman and Nave, 2002; Garman and Owen, 2006; Owen *et al.*, 2006). There is clear evidence that the results of radiation damage can lead to errors in the interpretation of the resulting structure (Dubnovitsky *et al.*, 2005; Grabolle *et al.*, 2006; Yano *et al.*, 2005). For example, in photosystem II, there are discrepancies between crystallographic studies and inconsistencies between the X-ray, electron paramagnetic resonance (EPR) and Fourier transform infra-red (FTIR) spectroscopic data (Yano *et al.*, 2005). In particular, the model of the Mn_4Ca complex from the X-ray data (Ferreira *et al.*, 2004) is 'highly unlikely' (Yano *et al.*, 2005). Under the dose conditions of the X-ray measurements the EXAFS spectrum changes significantly from one characteristic of a high-valent multinuclear oxo-bridged Mn_4Ca cluster to one that is characteristic of mononuclear hexacoordinated Mn(II) in solution (Yano *et al.*, 2005). It has been proposed that temperatures below 100 K might resolve the problem; however, photoreduction has been observed at temperatures as low as 20 K (Dubois *et al.*, 2006) and while radiation damage can be reduced, other undesirable features such as crystalline phase changes or crystal splitting may result (Hanson *et al.*, 2002).

Another metalloprotein example is the interpretation of the blue copper protein active sites as in *Pyrococcus furiosus* superoxide reductase (Clay *et al.*, 2002). Its oxidized copper(II) forms are considered strong oxidizing agents because their metal biosites are constrained to adopt a stereochemistry that is more appropriate for the reduced copper(I) state (Gray *et al.*, 2000). Based on a variety of spectroscopic studies, the oxidized copper forms are at least partially reduced during diffraction experiments (Clay *et al.*, 2002). Radiation-induced change in structures have been identified (Carugo and Carugo, 2005; Kouyama *et al.*, 2004; Roberts *et al.*, 2005; Weik *et al.*, 2002; Wilson *et al.*, 2003) and radiation damage can seriously affect the interpretation of structural data (Matsui *et al.*, 2002; van Thor *et al.*, 2002). Studies have shown specific non-uniform structural changes that progressively damage disulphide bridges (Burmeister, 2000; Helliwell, 1988; Ravelli and McSweeney, 2000; Weik *et al.*, 2000), cause decarboxylation of glutamate and aspartate residues, cause tryosine residues to lose their hydroxyl groups and cleave the carbon–sulphur bond of methionines (Burmeister, 2000; Ravelli and McSweeney, 2000; Weik *et al.*,

2000). The X-ray damage is not proportional to the X-ray absorption cross-section of the atoms making up the residues, and therefore must have a significant contribution from the reaction products of secondary electrons and the secondary electrons themselves.

Free-radical scavengers, styrene and ascorbate, have been used to mitigate X-ray damage in the case of lysozyme (Murray and Garman, 2002) with ascorbate showing practical utility. Nicotinic acid, oxidized glutathione and Ellman's reagent (DTNB) were used with three different protein systems. Significant protection was seen for nicotinic acid; DTNB almost doubled the useful lifetime of the crystal (Kauffmann *et al.*, 2006). Scavengers are widely used in electron spin resonance (ESR) spectroscopy as spin traps; for example 5,5-dimethylpyroline-*N*-oxide and 2,2,6,6-tetramethyl-4-piperidone. Such spin traps may have applications in crystallography (Murray and Garman, 2002). Other scavengers include ethanol, thiourea, (Blundell and Johnson, 1976) cysteine, glutathione, *t*-butanol, HEPES, Tris, glucose, and ethylene glycol. (Murray and Garman, 2002) There have been no systematic studies relating the scavenger performance to the macromolecule, biochemicals or the cryo-protectant. Specific scavengers may be better suited to specific cases, for other cases cocktails may be more effective than individual components. More recently, free-radical scavengers have been shown to be useful at ambient temperatures (Barker *et al.*, 2009).

Early studies by Henderson (1990) based on electron microscopy considerations and more recently studies with X-rays and crystal samples by Owen *et al.*, (2006) have given us a guide as to how much radiation dose our crystals can absorb, i.e. 2×10^7 Gy and 3×10^7 Gy respectively. Tools to calculate this dose based on sample and experimental properties are also available (Murray *et al.*, 2004; Murray *et al.*, 2005; Owen *et al.*, 2006) and have been integrated into control software for some synchrotron facilities (Gonzalez *et al.*, 2008). One should note that these are the maximum dose guidelines and it is better to use less rather than pushing the crystal to its theoretical limit.

Judicious use of wavelength (energy) has been proposed as a way of limiting radiation damage and sample heating with high photon energies (Helliwell and Fourme, 1983); (Helliwell, 1984; Helliwell, 1992). Further support for such beneficial possibilities comes from Arndt (1984) who considered the ratio both of the diffracted intensity per absorbed photon and per unit absorbed energy, and which is favourable on both counts, i.e. giving factors between 1.54 Å and 0.71 Å of 2.47 and 1.13, respectively. Unfortunately, with high photon energy and current detector technology comes the practical drawback of detector efficiency. Other practical disadvantages occur in terms of longer exposure times due to the λ^2 scattering efficiency effect (see Helliwell 1992 for example) and a steady increase in Compton background scattering (Gonzalez *et al.*, 1994). Geometrically, longer sample to detector distances are also needed for a fixed diffraction resolution limit and this may not be practicable if the sample is too mosaic in nature. The nature of the photon-energy dependence of the rate of radiation damage did not gain support from recent experimental results (Shimizu *et al.*, 2007; Weiss *et al.*, 2005). However, Weiss *et al.*, (2005) offer the caveat:- "However, some of the assumptions which had to be made to rationalize this observation are only roughly valid. For a better and a more quantitative assessment of what is really happening, it is obligatory that in future experiments a well calibrated pin diode is used to obtain accurate flux measurements, and that the size and the shape of the crystal as well as the surrounding mother liquor is recorded, and that the wavelength dependence of the detector response is taken into account." A feature of the Shimizu *et al.,* (2007) experiment is a very wide variation in diffraction resolution and $<I/\sigma(I)>$ at the edge of the pattern for the different wavelengths studied. Cianci, *et al.* (2008) examine the interdependence of wavelength, redundancy and dose in sulphur SAD experiments at six wavelengths between 0.979 and 2.29 Å wavelengths. A feature here was the freedom to use the tilted detector geometry at longer wavelengths to bring the measured diffraction data sets to a more homogeneous diffraction resolution to facilitate data set to data set comparisons. To extend to higher photon energies (shorter wavelengths than 0.979 Å) a further development is also vital, namely a more highly absorbing detector is needed at high photon energies (such as up to 40 keV) than, e.g. the CCD or image plate (see Helliwell (1992) p.206 for information on absorption curves going up to 80 keV photon energy). Nevertheless, Schiltz *et al.,* (1997) used the xenon K-edge and ESRF for SIRAS phasing of elastase as a test case. Also, Takeda *et al.* (2004) employed MAD phasing at the Xe and I K edge of hen egg-white lysozyme derivatives. Jakoncic *et al.,* (2006) successfully used 55 keV in a holmium MAD phasing study of lysozyme, using data recorded at ESRF ID15 and at NSLS with a MAR 345. These studies have an overall view to harnessing such absorption edges especially for radiation-labile samples whose structures are needed using MAD.

For consideration of microcrystals an interesting result is that of Nave (2005) who noted that for small enough crystals the finite track length of free electrons resulting from primary X-ray irradiation would be of such a length that these electrons might be ejected from the crystal without causing damage. Such an effect has been supported by microfocus experiments on the GA/CA-CAT beamline of the Advanced Photon Source showing a propagation of X-ray damage from the original exposure position (Fischetti and Sanishvili, personal communication).

14.3 Cryo-Cooling

Cryo-cooling is very effective at mitigating the effects of radiation damage, and has greatly simplified sample mounting and handling. There are many excellent practical references on the technique (Garman, 1999; Garman and Doublie, 2003; Garman and Schneider, 1997; Hope, 1990; Rodgers, 1997). The crystal is rapidly cooled from the temperature at which it was grown, down to a cryogenic temperature fast enough to produce an amorphous ice while avoiding the production of damaging crystalline ice (in this case ice, rather than solid water, is used to describe the largely aqueous mixture of components within the sample). The crystalline/amorphous ice transition temperature is approximately 140 K. For data collection the crystal is maintained at a temperature below this, typically 100 K. Crystalline ice is produced if the cooling is

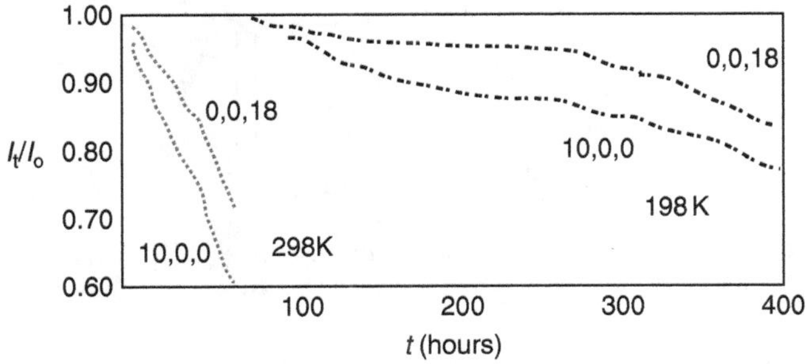

Fig. 14.2. The ratio of intensity at time t, I_t, versus the initial intensity, I_o, for two reference reflections plotted as a function of exposure time for a typical native and frozen crystal (with temperatures below the respective lines) (Haas and Rossmann, 1970). Used with the permission of the authors and the IUCr.

not rapid enough to prevent its formation. For the case of pure water, the cooling rate has to be on the order of 10^6 K s^{-1}(Lu and Liu, 2003; Peyridieu *et al.*, 1996) Cryo-protectants are typically used that extend this time allowing samples to be successfully cryo-cooled.

The potential usefulness of cryo-cooling was first noted by King (1958) and further experimentally demonstrated by Haas and Rossman (1970) who cooled lactate dehydrogenase crystals to 198 K and observed a decrease in the decay of reflection intensity over time, Figure 14.2. It has now become the *de facto* standard technique for data collection at the synchrotron and the majority of studies in home laboratories.

A major development that helped push cryo-crystallography into being the preferred technique was the development of crystal mounting in thin films suspended by a wire loop (Teng, 1990). This has been improved by use of nylon loops and more recently lithographic techniques producing rigid low-absorbing holders for samples (Thorne *et al.*, 2003). The complementary technology development was a reliable cryogenic system for the home laboratory (Cosier and Glazer, 1986) allowing extensive investigation and use of these techniques.

Cryo-cooling has revolutionized the success of macromolecular X-ray crystallography facilitating the ease of use of MAD methods and easing sampling handling. However, it can also be a cause of crystal degradation. Studies of macromolecular crystals at room temperature demonstrate a remarkable perfection (Bellamy *et al.*, 2000; Juers *et al.*, 2007; Ng *et al.*, 1997; Snell *et al.*, 2003; Vahedi-Faridi *et al.*, 2003). Cryo-cooling degrades this perfection through a simple increase in mosaicity due to thermal contraction (Vahedi-Faridi *et al.*, 2003) or with inappropriate cryo-protectant conditions, Figure 14.3. The cryo-protectant may have a significantly greater osmotic pressure than the crystal causing osmotic shock and cracking manifesting as several lattices in the diffraction pattern, an increased mosaicity, spot streaking or reduced diffraction resolution. If the cryo-protectant alters the solubility of conditions the sample may start to dissolve. Small-molecule cryo-protectants enter

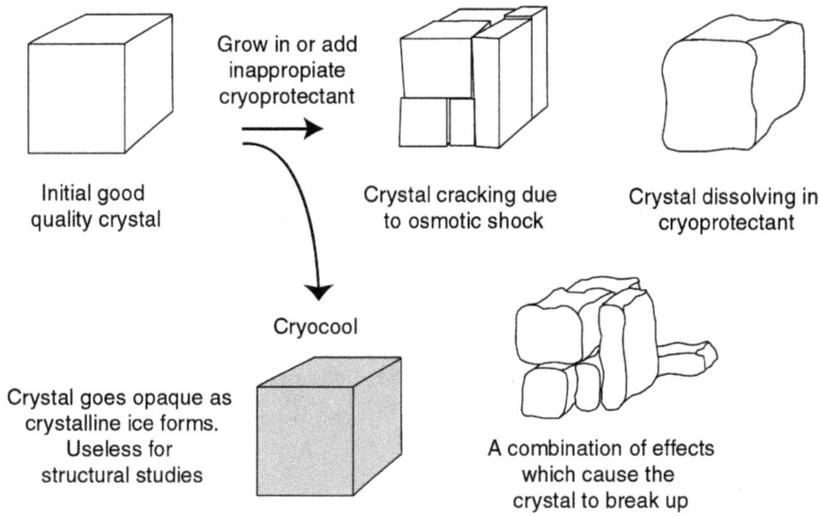

Fig. 14.3. A schematic portrayal of degradation of the crystal quality from poor cryo-protectant or cryo-cooling practice.

the crystal lattice rapidly and may destroy short-range order leading to a loss in diffraction resolution. In some cases the cryo-protectant may preferentially bind to the macromolecule displacing water and disrupting the lattice. A practical step to avoid this is crystal growth including a small concentration of the cryo-protectant of interest.

There are a number of studies on different cryo-protectants that can be used and the appropriate concentrations needed (Berejnov *et al.*, 2006; Garman and Mitchell, 1996; Kempkes *et al.*, 2008; McFerrin and Snell, 2002). Equally important is the method that the cryo-protectant is introduced to the crystal samples (Garman and Doublie, 2003) and the osmolarity of the cryo-protectant solution in comparison to that the crystal was grown in (Garman, 1999).

The success of cryo-cooling can be gauged visually. If the sample is clear on cooling, i.e. no visible ice is present, then it is likely to have been successfully cryo-cooled into an amorphous state (Chinte *et al.*, 2005; McFerrin and Snell, 2002). This is the first stage of cryo-cooling optimization. Like crystallization, the cryo-protectant stage also requires optimization and careful investigation of the cryo-protectant concentration or type can greatly improve the results. For example, Mitchell and Garman (1994) note that a change in glycerol concentration can significantly reduce the mosaic spread and improve the diffraction limit, Figure 14.4.

Annealing is another technique that can be employed to improve crystal mosaicity (Ellis *et al.*, 2002; Hanson *et al.*, 2003a; Hanson *et al.*, 2003b; Harp *et al.*, 1999; Harp *et al.*, 1998; Kriminski *et al.*, 2002; Yeh and Hol, 1998). It would seem that this is most successful when the initial cryo-protectant concentration is higher than that

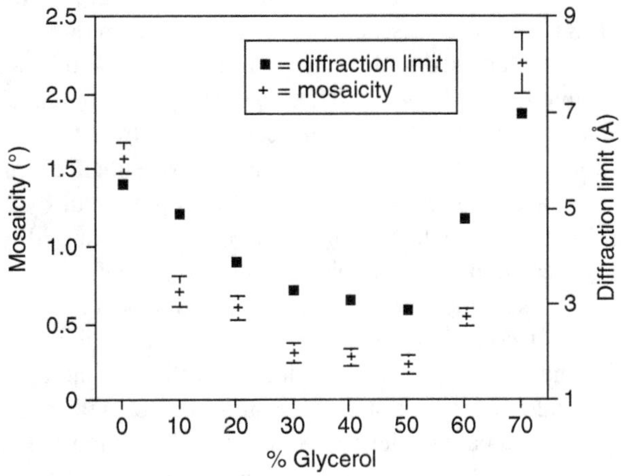

Fig. 14.4. Mosaicity and diffraction limit plotted as a function of glycerol concentration (Mitchell and Garman, 1994). Used with permission of the authors and the IUCr.

needed for successful cryo-protection (Juers and Matthews, 2004b). Annealing allows water transport and changing of the cryo-protectant concentration within the crystal. In doing so, this alters the bulk solvent thermal contraction and in turn compensates for that of the lattice (Juers and Matthews, 2004b). There are many excellent reviews on cryo-crystallography (Garman and Doublie, 2003; Garman and Owen, 2006; Garman and Schneider, 1997; Hope, 1988; Juers and Matthews, 2004a; Mitchell and Garman, 1994; Parkin and Hope, 1998; Pflugrath, 2004; Rodgers, 1994; Rodgers, 1997; Walker *et al.*, 1998). All detail the correct technique and practicing of that technique.

14.4 Dehydration and Humidity Control

Macromolecular crystals have a high solvent content, from ~30–70% (Matthews, 1968; Kantardjieff and Rupp, 2003). For data collection at ambient temperatures the crystals are typically mounted in an environment that maintains their humidity. An early method to do this was the use of glass capillaries for X-ray crystallographic studies of pepsin (Bernal and Crowfoot, 1934). As early as 1946 Perutz (1946) used glass capillaries to help control the humidity of haemoglobin crystals. Huxley and Kendrew (1953) observed discontinuous changes in lattice constants, distinct types of lattice and a reversible effect, even at the fully dry stage. Changes appeared to occur at defined vapour pressures. Pickford *et al.* (1993) built on these results developing a humidity cell that allowed the relative humidity environment to be controlled within the capillary during X-ray data collection.

Ribonuclease and lysozyme exhibited similar changes on varying the humidity (Salunke *et al.*, 1984). Ribonuclease showed distinct drops in unit-cell volume as the relative humidity was decreased with a sharp drop at 95–93% followed by a plateau then a gradual decline from 88% to 75%. Lysozyme showed a gradual decrease from 100 to 95% with a more rapid decline in unit-cell volume from 95% to 90% followed by a gradual further decrease. There was no noticeable enhancement of diffraction properties in either case. Lysozyme was studied in more detail by Morozov *et al.*, (1985) showing a metastable state occuring between 60–70% humidity. Morozov and Morozova (1981) measured the Young's modulus (essentially stiffness) of haemoglobin, myoglobin, lysozyme and cross-linked lysozyme crystals. The humidity decrease resulted in a considerable qualitative increase of the Young's modulus. The humidity dependence of Young's modulus may be explained by the fact that neighbouring molecules in the crystal come into contact with each other through relatively small surface areas. Under the action of dehydration the macromolecular crystal has a water activity lower than its mother liquor and is therefore subjected to internal pressure. The molecules making up the crystal must come closer together, increasing surface contact areas. Morozov *et al.*, (1985) calculate the pressure according to the Thompson (Kelvin) equation;

$$P_L = -(RT/V_e)\ln(A/A_{cr}) \qquad (14.1)$$

where A/A_{cr} is the relative humidity, R is the gas constant, T is the absolute temperature and V_e is the molar volume of water. They observe good agreement with the calculated pressure, 0.035 GN m^{-2} at 70% humidity, resulting in a 3.5% compression, and the observed shrinkage of the crystal.

Ribonuclease A, lysozyme, insulin and concanavalin A were studied by Salunke *et al.*, (1985). Ribonuclease A was grown from acetone in tris buffer, aqueous ethanol and MPD solution in phosphate buffer. In the acetone case the initial solvent content was 40%. A transition occurred in the 95–93% range on reducing the relative humidity with the solvent content dropping to 36% and the unit-cell volume decreasing ~6%. For the ribonuclease A crystal grown from aqueous ethanol a transition occurred between 93 and 90% and again at 88 to 84%. The solvent content dropped from 44% to 39% and then 31% with the unit cell shrinking by ~5% and ~18%, respectively. The sample grown in MPD showed no transition in the relative humidity range of 100 to 75%. Lysozyme showed similar trends but insulin and concanavalin A did not. The diffraction patterns were collected under similar conditions (Salunke, 2007) and interestingly, in the lysozyme case, there is a significant enhancement in the overall diffraction intensity at 93% in comparison to 100% relative humidity (Figure 14.5).

A key experiment prompting humidity control as a means to increase crystal quality came with the study of HIV-1 reverse transcriptase (Esnouf *et al.*, 1998). The initial 'control' of relative humidity in this case came not from a planned experiment but from the fortuitous observation of increases and decreases in the diffraction limit of a crystal during data collection. On dismounting the crystal, it was noticed that the crystal was no longer surrounded by mother liquor and there was a

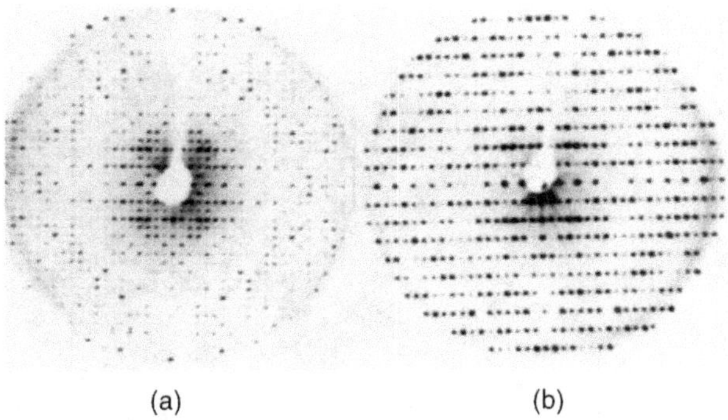

Fig. 14.5. 0*kl* 15 precession photographs taken from monoclinic lysozyme at (a) 100% and (b) 93% relative humidity (Salunke *et al.*, 1985). Used with the permission of the authors and the IUCr.

small crack in the capillary tube wall suggesting unplanned dehydration. Deliberate dehydration was carried out by moving the crystals from their initial mother liquor, containing PEG 3400, into buffered solutions of increasing PEG 3400 concentration. This reduced the overall solvent concentration from 56% to 48% and led to an increase of diffraction resolution from 3.7 Å to 2.2 Å; an almost five-fold increase in the number of observed reflections (Esnouf *et al.*, 1998; Ren *et al.*, 1995). Similar

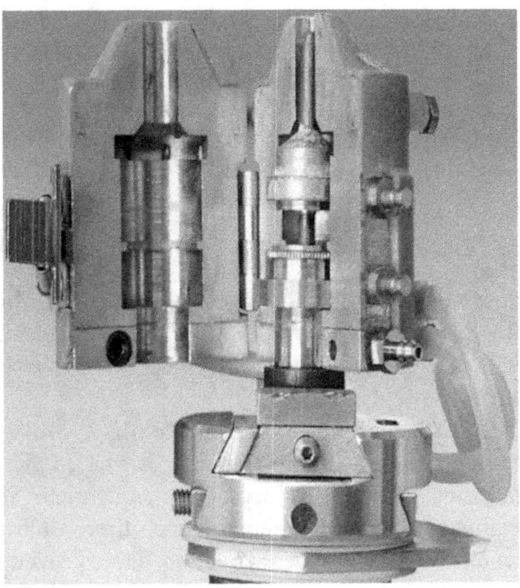

Fig. 14.6. The free mounting system (photograph courtesy of Proteros biostructures GmbH).

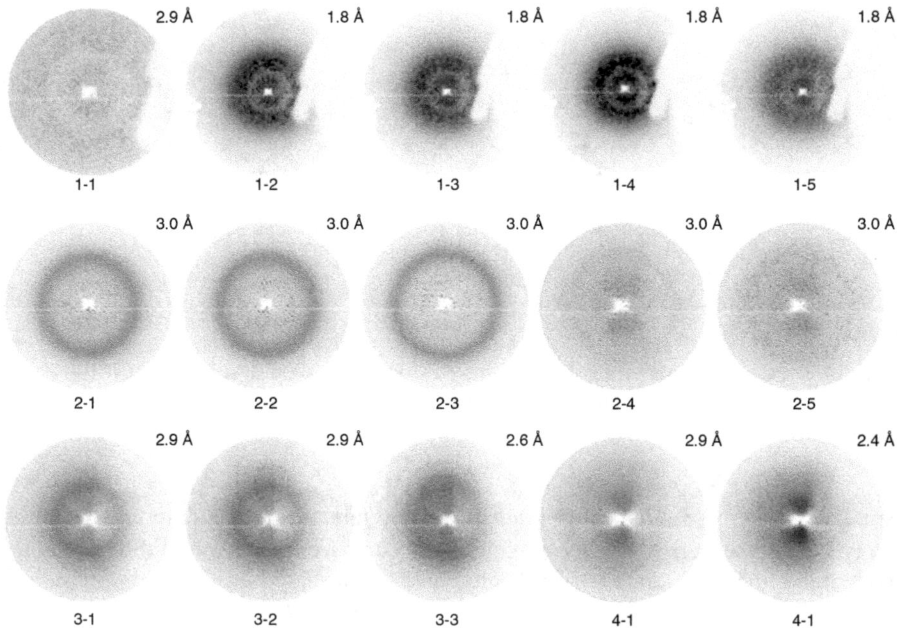

Fig. 14.7. Crystal transformation examples from Kiefersauer *et al.*, (2000). Diffraction images showing CODH transformation: (1-1) to (1-5), during variation of humidity (96%, 93%, 84%, 81%, and 90%, respectively). Oxidase transformation: (2-1) to (2-5). The diffraction images (2-1) and (2-2) were exposed for 3 min. Then the crystal was frozen, (2-3). The crystal was mounted on the pin of a micropipette, (2-4). The humidity was adjusted to 99% then the humidity was reduced by 2%, (2-5). DP4 transformation: (3-1) to (3-3). The adjusted humidity was 93.5%. The humidity was then reduced to 83% with a gradient of 0.5% (20 s)$^{-1}$, (3-2). (3-3) The crystal was then shock frozen. NAP transformation: (4-1) to (4-2). The crystal was mounted on the pin of a micropipette, (4-1). The humidity was adjusted to 97%. After dehydration to 95%, the humidity was then increased to 99.5%, (4-2). Used with the permission of the authors and the IUCr.

dramatic results have been seen for other crystals (Heras and Martin, 2005). However, dehydration by this method does not always increase the diffraction resolution (Iwasaki *et al.*, 1999) but can produce phase transitions that may help understand function (Weiss and Hilgenfeld, 1999).

Samygina *et al.*, (2000) introduced reversible flash cooling to dehydrating PPase crystals improving the resolution from 1.8 Å to 1.15 Å. They X-rayed the cryo-cooled crystal, then warmed the crystal in sequentially more concentrated salt solutions repeating the X-ray examination each time. Both the salt and cryoprotectant had a beneficial effect on the diffraction. Time proved to be a significant factor with the improvements seen with short (3–6 s) soaking over longer (15 s) soaking times.

Cryo-crystallography developments introduced the method of loop mounting (Teng, 1990), which can also be accomplished at room temperature by enclosing the loop or other mount within a capillary (Kalinin *et al.*, 2005; Mac Sweeney and D'Arcy, 2003). An alternative strategy is to keep the mounted crystal in a humid environment by passing a stream of humid gas over the crystal. A system to do this has been developed by Kiefersauer *et al.*, (2000) and commercialized as the free mounting system, Figure 14.6. The system has been applied to a number of macromolecules, Figure 14.7 shows examples of diffraction patterns from CO dehydrogenase (CODH), ba_3-cytochrome c oxidase, dipeptidyl peptidase IV and periplasmic nitrate reductase (Kiefersauer *et al.*, 2000).

The free mounting system has been used for a number of studies (Bowler *et al.*, 2006; Dobbek *et al.*, 2002; Engel *et al.*, 2006; Engel *et al.*, 2003; Estebanez-Perpina *et al.*, 2000; Henrich *et al.*, 2003; Koch *et al.*, 2004; Kyrieleis *et al.*, 2005; Saric *et al.*, 2007). A similar system has been developed by Sjogren *et al.*, (2002) where they have also proposed its use in the anaerobic manipulation of crystals (through the use of different gasses), causing small pH changes, or the introduction of gaseous or volatile ingredients and reactants into the crystal.

14.5 Chapter Summary

The growing experience of handling crystals, not least cryo-cooling, annealing and dehydration, as well as an ever-growing capability of technology, apparatus and technique continue to march forward hand in hand to allow ever more complicated structural biology studies. Thus, the Protein Data Bank at the time of writing this book comprises over 50 000 depositions, 90% of which come from crystallography. It is a major achievement of this field of science and its worldwide practitioners. Crystals do degrade under X-ray diffraction study but we have learnt both to manipulate the improvement of their perfection and know better the bounds and risks of how we degrade them in our studies.

15

Unusual diffraction geometries

X-ray diffraction physics applications of protein crystals such as in Laue geometry, the large-angle oscillation technique and forming polychromatic profiles across diffraction spots are examples of unusual diffraction geometries recently summarized in Helliwell (2005). The Laue geometry, of these three less usual measuring geometries, is now well established for both very fast 'pump probe' time-resolved crystallography of light-sensitive proteins and neutron macromolecular crystallography data collection.

The situations of Laue and the large oscillation technique (LOT) (Weisgerber and Helliwell, 1993b), with their more crowded spot patterns and the closer approach of neighbouring diffraction spots puts a premium on diffraction spot sizes being kept to a minimum. Some flexibility is still possible in the software through spot profile integration and spatial deconvolvution of spots that might partially overlap.

Figure 15.1 shows a neutron Laue diffraction pattern for a concanavalin A-glucoside complex under typical experimental conditions. This is amongst the largest proteins, at 50 kDa protein in the asymmetric unit, studied so far with neutrons (Gilboa *et al.*, 2001). The protein crystal size actually used was very large at 3 mm diameter being 'icosahedral cubes'; note that there were larger crystals available, i.e. that grew to 5 mm diameter but would have led to severe spot to spot overlap in the neutron Laue pattern and so the smaller crystal had to be used.

In the X-ray case, a 'toast-rack' detector scheme was used for photographic Laue data collection on macromolecular crystals, Figure 15.2. This is a novel geometry for the stacking of photographic film or image plate/ storage phosphors (IP/SPs) (Helliwell, 1991). The 'toast-rack' exploits the fact that in a Laue pattern there is a strong correlation of stimulating wavelength with Bragg angle; moreover, the densest parts of the pattern occur at a Bragg angle of :

$$\theta_{\text{dense}} = \sin^{-1}(\lambda_{\min}/2d_{\min}) \tag{15.1}$$

where $\lambda_{\min}$ is the minimum wavelength in the illuminating X-ray wavelength bandpass and $d_{\min}$ is the diffraction resolution limit.

By placing several films (e.g. up to 10) parallel to one another but spaced 1 cm apart (Figure 15.2 top) the dense short-wavelength parts of the pattern can penetrate to the rear film and expand to fill the available area. In the case of the more strongly absorbing and larger IP/SPs two of these can be placed say 40 cm apart to allow the rear IP/SP to

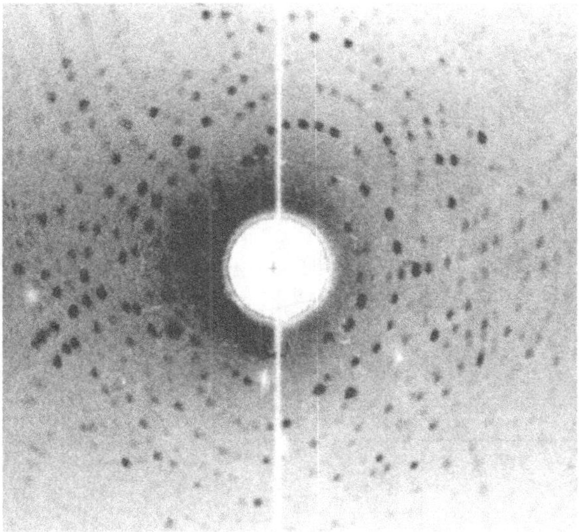

Fig. 15.1. Example of a congested neutron Laue crystal diffraction pattern with larger than usual diffraction spots due to the large crystal size ($\sim$3 mm) used for concanavalin A in crystal space group $I2_13$, $a = 167.8$ Å. This was recorded at the Institut Laue Langevin Laue Diffractometer (LADI-I) for biological crystallography. From Gilboa *et al.*, (2001) with the permission of the authors and the IUCr.

pick up the short-wavelength densest part of the pattern. Tests have been made with both film and storage phosphor, and that shows that good spatial resolution of spots is achieved with good background on the rear film/SP (see Figure 15.2 bottom). The detector scheme is well suited to the larger unit cell protein and virus Laue patterns. The limitation is that neither film nor image plate/storage phosphor are electronic nor lend themselves therefore to automatic, immediately digitized, data collection. An electronic scheme would still be possible by employing a small-angle X-ray scattering/ wide-angle X-ray scattering (SAXS/WAXS) style of setup, i.e. a close-in WAXS detector with a central hole would let the low-angle diffraction pass through to a detector set at a longer distance. The Laue spots recorded on the long-distance detector would obviously benefit from a lower mosaicity crystal sample. The spot size in the Laue geometry is dominated by a radial extension given by:

$$\text{radial} = D \sin(2\eta) \sec^2 2\theta \tag{15.2}$$

where D is the distance to the flat detector plane surface set perpendicular to the X-ray beam and θ is the Bragg angle of the given Laue spot, and assumes that the beam convergence angle is smaller than the mosaic spread of the sample crystal 'η'. Full details can be found in Andrews *et al.*, (1987).

The principle for a large-angle oscillation technique (LOT) and example X-ray diffraction pattern (Weisgerber and Helliwell, 1993b) is shown in Figure 15.3.

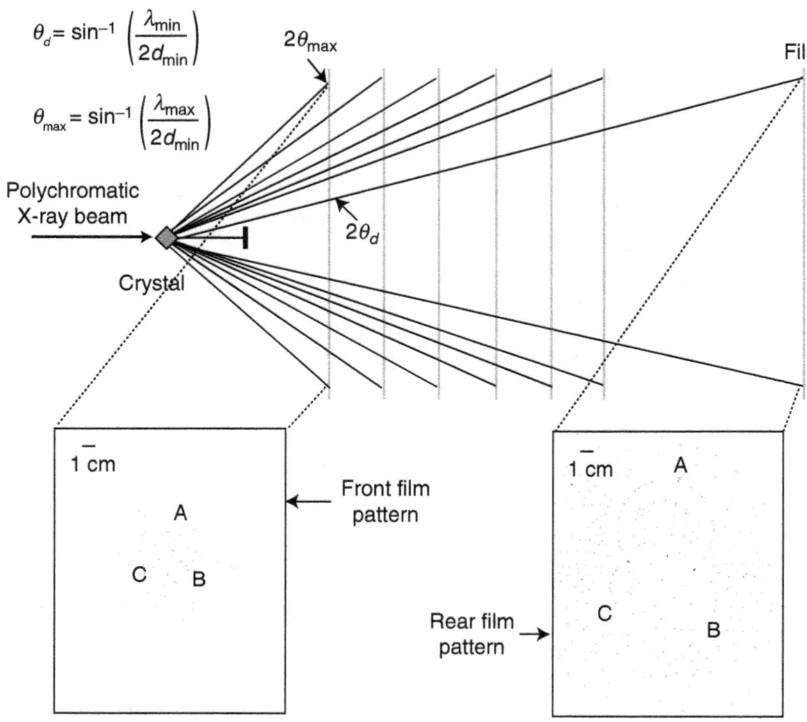

Fig. 15.2. Toast-rack detection scheme (top) principle illustrated with a set of parallel spaced films; (bottom) test experiments done at CHESS with two storage phosphors spaced 40 cm apart (Legrand, Harrop and Helliwell, unpublished). Nodal spots marked A, B, and C, show that the nearest-neighbour singles are resolved on the back phosphor, whereas they were too close together on the front phosphor.

Randomly set crystal diffraction pattern examples are also given in the same reference. These LOT patterns lend themselves to doing very fast sample rotations for time-resolved applications, i.e. 'shutterless' data collection. Whereas initially demonstrated with photographic film this diffraction geometry would still apply to 'pixel electronic detectors'. Obviously electronic shuttering may be more convenient than simply carrying on rotating. Nevertheless, these patterns demonstrate that the usual statement of restricted rotation angle per diffraction image, e.g. 1 degrees maximum typically being employed, is not necessarily a required experimental restriction.

Thirdly, where diffraction spot intensities are weak such as at high resolution an angularly sharp reflecting range allows the strategy of using fine phi slicing to improve data quality. Since protein crystals have been shown to reach 'silicon-level-quality' (Fourme *et al.*, 1995; Helliwell, 1988; Snell *et al.*, 1995b) super-fine-phi slicing to extend the diffraction resolution limit has been investigated (Borgstahl *et al.*, 2001); see Figure 15.4. Detectors such as the pixel detector optimally can exploit such very narrow reflecting ranges.

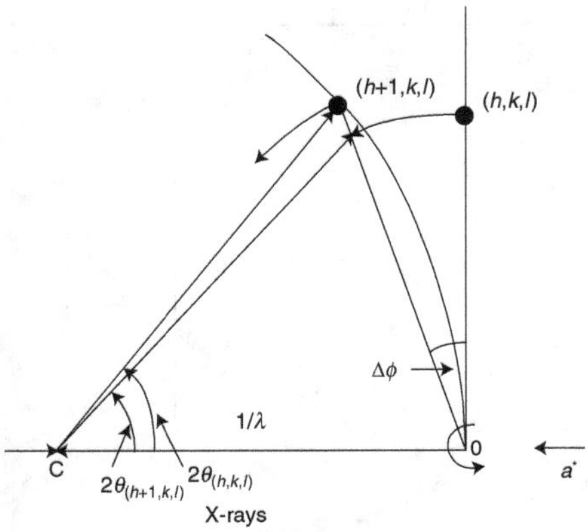

Fig. 15.3. The case of the large-angle oscillation technique (LOT) diffraction (from Weisgerber and Helliwell (1993b). Reproduced with permission of the Royal Society of Chemistry.). (a) The principle behind LOT; the rotation of the crystal beyond the typical $\Delta\varphi_{max}=d_{min}/$ maximum unit-cell dimension is not in fact limiting as the Bragg angles of these two reciprocal lattice points (h, k, l and $h+1$, k, l) are slightly different. (b) LOT practice; a 90° rotation X-ray diffraction photograph recorded at SRS station 9.5, wavelength ~1 Å from concanavalin A saccharide-free crystal form space group I222, unit-cell dimensions $a = 88$, $b = 86$, $c = 62$ Å, diffraction resolution at the edge is 5 Å (higher resolution, $d_{min} = 3$ Å, randomly set crystal diffraction pattern examples are also given in Weisgerber and Helliwell, (1993b).

The more specialized diffraction geometry involving the Laue method has had important applications in single-bunch synchrotron X-ray time-resolved and neutron macromolecular crystallography structural studies.

The wish to understand biological and chemical mechanism based on 3D structure change and dynamics took off in spectacular fashion when it was realized in the 1980s that with the new SR 3rd-generation sources if one focused the ultra-bright white X-ray spectrum emission onto a single protein crystal one could measure data with a single sub-nanosecond pulse of X-rays from a single bunch in the SR machine. Thus the time resolution for measurement was in the sub-nanosecond range. This required a return to the Laue method including the overturning of misconceptions about the method with new theoretical treatments of energy and spatial overlaps as well as new software and analyses methods. The pump-probe method, familiar to physical chemist spectroscopists, was thereby applicable to the protein crystal state. Then, first at ESRF and then at APS, such experiments were performed with carbon monoxy myoglobin 'CO-Mb' (Srajer *et al.*, 2001; Srajer *et al.*, 1996) and photoactive yellow protein 'PYP' (Ihee *et al.*, 2005; Kim *et al.*, 2004; Ren *et al.*, 2001). There are complementary approaches namely of (i) freeze trapping structural states, for slower kinetic processes in the crystal applicable to, e.g. diffusion of substrates to enzymes and (ii) studying protein inhibitor complexes, to yield several static crystal structures

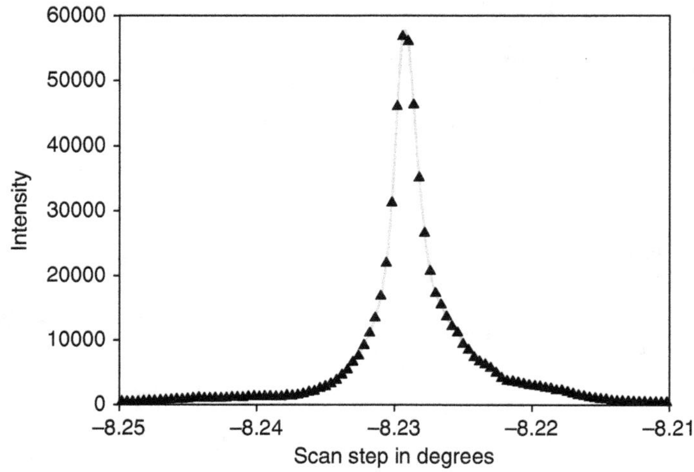

Fig. 15.4. Sharp mosaicity of a protein crystal illustrated by a high-resolution reflection from a near-perfect lysozyme protein crystal. See Snell *et al.*, (1995b).

from which molecular movies, Disney cartoon style, can be created and finally (iii) molecular-dynamics computer prediction of structure change. Cases (i) and (ii) involve measured data via the X-ray monochromatic rotating-crystal method. Overall, the whole field of macromolecular structure and dynamics has expanded considerably in the last 15 years.

The determination of the hydrogen and hydration details in a biological macromolecule has been revitalized in the last 10 years. Studies of enzyme mechanism on the one hand and ligand receptor recognition on the other being prime examples of such applications. A range of apparatus has been developed involving use of monochromatic and Laue geometry. The weak fluxes of neutron sources place a premium on harnessing as much of the emitted spectrum as possible, hence the interest in Laue methods at both reactor and spallation sources. A number of reviews of this expanding field are available (Blakeley *et al.*, 2004; Blakeley *et al.*, 2008; Niimura and Bau, 2008). It is important to emphasize that there is a complementarity of X-ray and neutron techniques, i.e. that are optimally used in harness together to study protonation states of amino acids like Asp, Glu and His and of mechanistic/catalytic bound waters either as H_2O, OH^- or H_3O^+.

15.1 Chapter Summary

Unusual diffraction geometries may seem a curiosity but may stimulate novel avenues of application most notably in the way the Laue method has with time-resolved X-ray studies and data collection with neutrons. Not least they illustrate a diversity of diffraction measuring possibilities.

16

Making the most of difficult crystals – beamline and detector optimization

16.1 Introduction

There are many situations where the quality of the crystal cannot be improved. In these cases, and for all data collection, we should be aware of strategies to make best use of the instrumentation available. Geometry is one aspect, optimizing the beam to the sample is another. In this chapter we cover both of these but also touch on other experimental approaches that can also help. A more detailed treatment than is given here can be found for example in International Tables Volume F in various chapters (Editors E. Arnold and M. G. Rossmann). Here, we treat also the special case of microcrystals, i.e. in the 1 to 20 micrometre size range, which we cover in detail. Unusual diffraction geometries were described in Chapter 15.

16.2 Geometry and End-station Instrumentation

In producing the best-quality diffraction data flexibility is a key requirement of a beamline. An equally important requirement is minimizing the noise and maximizing the signal. In terms of flexibility the two geometries that can be altered to help an experiment are detector distance and offset. Experimentally the oscillation angle is another factor that can aid the best data collection.

In the case of detector distance, this can be optimized to the unit cell size and/or the wavelength used. An example of this is shown for Photosystem II in Figure 16.1. The crystal had a largest unit cell of 357 Å that, even to collect data to low resolution, required a crystal to detector distance of 500 mm.

Detector offset can be used to achieve a similar separation of reflections but at the expense of limiting the area of the diffraction pattern captured and necessitating significantly more images for overall completeness of the data. This can be achieved in two ways, either by rotating the detector about the 2θ-axis, which is in the vertical plane in the synchrotron case due to the horizontal polarization of the beam, or by translating the detector vertically. In the latter case more smearing of the spots can

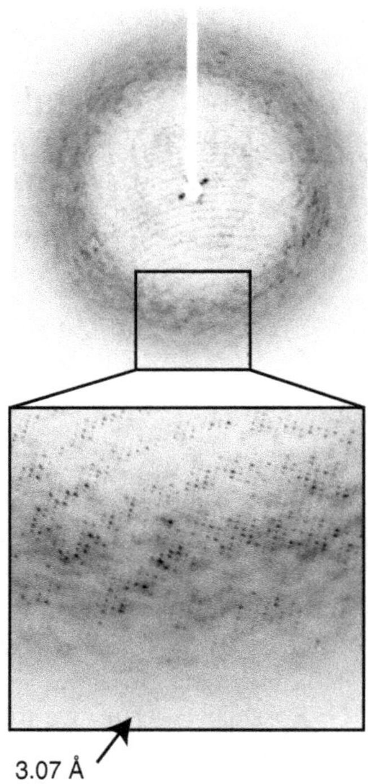

Fig. 16.1. An example of a very long unit cell edge(s) (longest 357 Å) and the necessary distance detector setting (500 mm) for red algae Photosystem II from Adachi *et al.*, (2009) with permission of the authors and Elsevier.

take place as they impinge on the detector at increasingly oblique angles. The 2θ-axis is illustrated in Figure 16.2, which also shows how crowded a typical beamline end-station can become when all the typical experimental equipment is present.

There can be cases where the cell edges are all large as with a virus crystal but also, e.g. human STAT-1, where one cell edge is very long in comparison to the others (Figure 16.3).

16.3 Signal and Noise Considerations

Reflection intensity signal and underlying noise are two other key considerations. Obviously one wants to maximize the former and minimize the latter. There are general principles to do with systematic errors and random errors, and then specific challenges associated with particular types of data-collection experiment. From the source and beamline optics we seek stability, i.e. minimal intensity fluctuations over time. From the crystal perspective we seek accurate and reproducible positioning of the sample. In the case of data completeness we want the ability to orient the crystal

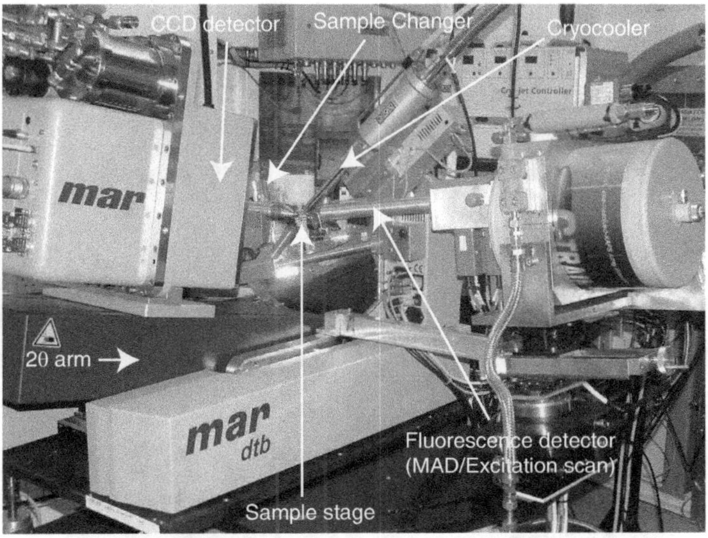

Fig. 16.2. General view of an experimental end station that allowed tilting of the detector of up to 30 degrees (the Marreseach desktop beamline, Mar225 CCD). Also seen is an ORTEC C-TRAIN-04 detector for X-ray fluorescence spectral signals to be measured (Cianci *et al.*, 2005a) with permission of the authors and the IUCr.

such that we can collect 100% of the data with high redundancy and if necessary pay particular attention to symmetry-related pairs. In the case of weak anomalous differences we would like to measure them as close together in time as possible. So, it can be possible to specifically orient a crystal to have the Friedel differences measured basically simultaneously with time and yield anomalous differences that are thereby not sensitive to time-dependent fluctuations (see, e.g. Einspahr *et al.* (1985) and Nieh and Helliwell (1995)) and Figure 16.4.

From the detector we would like: as high a sensitivity as possible, with small X-ray losses through the front entrance window, consistent with adequate dynamic range, a stable response, as well as good overall and local count rates. Careful calibration of any spatial distortion or any intensity response non-linearity may be needed, and checked for any changes with time, e.g. days to weeks. In addition, the detector quantum efficiency (DQE) is a very effective measure of a detector's signal versus noise performance and is the ratio:

$$(S/N)_{out}/(S/N)_{in} \qquad (16.1)$$

A plot of DQE versus photon signal readily reveals weaknesses in any given detector device $(S/N)_{out}$ measurement performance (Lewis, 1994).

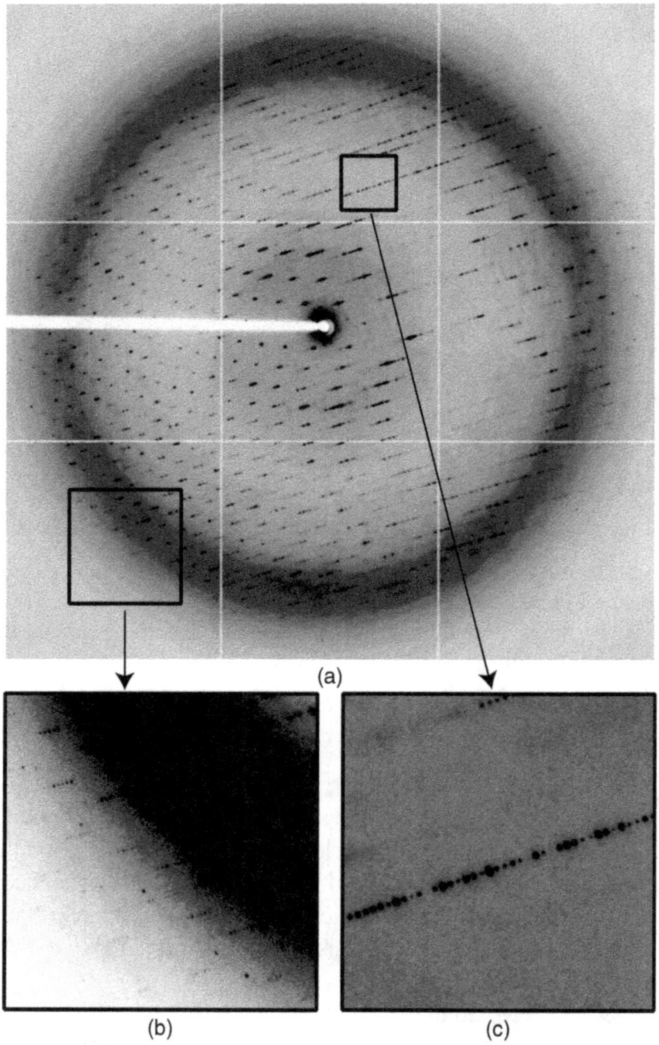

Fig. 16.3. (a) X-ray diffraction image of a Na_2OsCl_6-derivatized crystal of unphosphorylated human STAT1 (1–683) in complex with the phosphopeptide; (b) and (c) magnified views of the boxed regions in (a). The crystal belonged to space group $P6_122$, with unit-cell parameters $a = b = 102.6$, $c = 646.5$ Å. From Mao and Chen (2005) with permission of the authors and the IUCr.

We can separate the signal and noise and consider each individually. In general, the reflection signal from the crystal can be enhanced and optimally measured by a detector by:

1. Having as large a crystal volume in the beam as possible, up to a limit set by crystal absorption of the X-rays and that depends on the wavelength via the

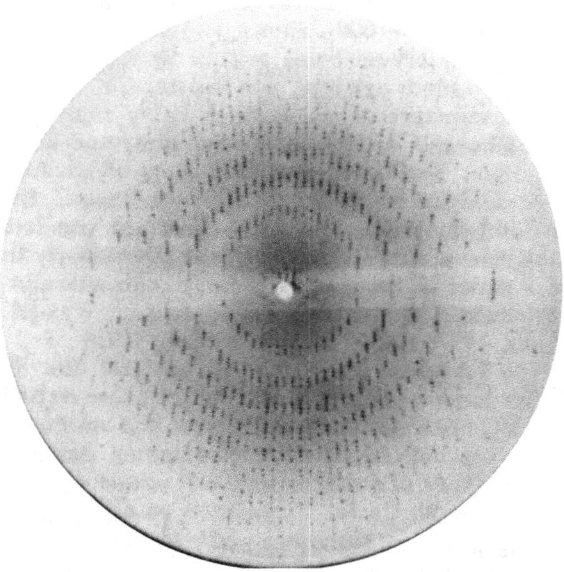

Fig. 16.4. A carefully aligned crystal allows anomalous differences to be measured from Friedel pairs stimulated close together in time. In this case, pea lectin, the wavelength is set at 1.896 Å at the Mn K edge and anomalous differences are measured between Friedel pairs (left and right of the vertical axis). From Einspahr *et al.* (1985) with permission from the authors and the IUCr.

linear absorption coefficient of the sample and the thickness of the crystal in the given beam-path direction; an extreme example of sample absorption is the case of ferritin that contains more than a thousand iron atoms within a protein shell!

2. Increasing the incident X-ray intensity; there are thermal and radiation-damage limits here (see Chapter 14).
3. Increasing the exposure time; for a detector like a CCD with a rising noise floor with time the detector dynamic range can be 'eaten away' and make the strong reflections saturate the device; to mitigate against this both 'slow' and 'quick' pass data sets are recorded, which in effect extends the dynamic range measurable; in extreme cases such as ultrahigh-resolution data collection even a medium-pass data set is recorded as well.
4. Increasing the wavelength as the Signal from the crystal sample is proportional to λ^2, e.g. using up to 2 Å wavelength; the beam paths to the detector can then suffer air absorption but a fairly simple measure to reduce such absorption can be taken by using a helium-filled cone to replace the air; for longer wavelengths a more technically complex approach would be to place the diffractometer in vacuum (Ohsumi *et al.*, 1991).

5. A horizontal rotation axis places the geometric 'blind region' direction to be favorably coincident with the polarization attenuation effect of the scattered X-ray beams in the horizontal direction especially important for the horizontally polarized X-ray beam case of synchrotron radiation.

The background noise under a reflection signal from the crystal can be reduced optimally by:

1. Having the X-ray beam size cross-section matched to the diffracting sample cross-section, thus minimizing any sample holder material in the X-ray beam (cryo-loop frozen mother liquor or glass for capillary, usually room temperature, data collection);
2. Having a well-collimated X-ray beam, i.e. as parallel and non-diverging a beam as possible thus reducing any interaction with air (or helium gas) on the path to the detector as much as possible and, for a detector that may introduce noise into the measurement of a reflection signal intensity, minimize the spot size (Helliwell *et al.*, 1982); a further advantage is that with a larger detector it can be placed further away than a smaller device and reflection S/N approximately improves as the inverse square law with distance if the spot sizes remain fixed (the approximation being that some of the background features are to some degree also collimated such as air scatter and different types of diffuse scatter from the sample crystal);
3. Having the X-ray direct beam stop as close to the sample as possible;
4. Avoiding any upstream sources of X-ray background that may arise, e.g. due to poor lead shielding placement and direct 'lines-of-sight' to the upstream optics or source itself;
5. Using rotation ranges for a given diffraction reflection that are no wider, ideally less, than the intrinsic rocking curve and that is known as 'fine phi slicing' measuring mode; since room-temperature crystals can be perfect or near-perfect, rocking ranges need only be of the order of 0.001 degrees! Frozen crystals are more typically with a reflection rocking curve in the range of 0.2 degrees or larger; high mosaicity values of 1, even 2 degrees or more, do occur and pose special problems in data collection and processing. Microgravity and gel-media crystal growth have been shown to improve such mosaic crystals to be more acceptably perfect (see Chapter 8);
6. Different software processing packages do offer differing features although today there is a considerable degree of convergence on algorithms used such as reflection intensity profile fitting; current status of such software upgrading includes dealing better with twinned or textured crystals or dealing with several tiny crystals in the beam simultaneously.

With these approaches, the S/N for each reflection and the overall diffraction resolution limit can be optimized as much as possible. The community, most widely accepted, criterion for the measurable edge of the diffraction pattern is $<I/\sigma(I)>$ crossing a value of 2.0.

There are several potential sources of systematic error. Assuming that the various possible systematic effects (distortions and non-linearities) of the detector are properly calibrated these include effects from absorption in the sample itself. The primary-beam and secondary-beam path lengths passing through an irregularly shaped crystal in its various orientations are then the source of errors between otherwise equivalent reflection intensities. It is sometimes stated that high redundancy of measurements irons out such errors but this is not the case. Clearly shorter wavelength and/or smaller crystal can be harnessed to reduce such effects, but this is not always viable. A good practice to adopt is that from chemical crystallography, which is to calculate the theoretical 'T_{max}/T_{min}' for the maximum and minimum transmitted intensity that you might expect knowing your preferred wavelength, likely crystal size and shape and, perhaps, unusual chemical composition as with the ferritin mentioned above. Data-collection plans may need to be amended!

Thus, if we can minimize intensity measurement error, be it systematic or random noise, we can maximize the quality of the resulting data, the ease of determining the structure and finally the quality of the determined structure.

16.4 Very Small Crystal Volume 'Microcrystals'

An increasing number of structural biology research projects seek to move forward with ever-smaller crystals. An informal survey of the pre-ponderance of such projects is mentioned at the end of Chapter 17. Why crystals won't grow to a 'standard size' after nucleation is an intriguing question. It may be thought that tiny crystals are not well formed, as chemists say *"the sample is probably only a 'sludge'"*. However, gazing into 'protein microcrystal slurries' can have its rewards if it yields up microcrystals.

The ever-increasing X-ray spectral brightness of synchrotron sources is obviously especially relevant to this topic. The term 'X-ray brilliance' is sometimes used, more so in the past than today (see Mills *et al.* (2005) for a full and proper definition of nomenclature terms). Either way this key source parameter is expressed in units:

photons s^{-1} mm^{-2} $mrad^{-2}$ $(0.1\%\ \text{bandwidth})^{-1}$

and more pertinently for the topic of microcrystals:

photons s^{-1} micrometre^{-2} microrad^{-2} $(0.1\%\ \text{bandwidth})^{-1}$

The history of the development of synchrotron-radiation microcrystallography started really with chemical crystallography cases as such samples were more readily within reach of the second-generation source intensities. Thus, studies on CaF_2 by Rieck *et al.* (1988) for example were pioneering and led on to structure determinations of challenging cases in chemical crystallography such as piperazine silicate and, which indicated the scope of possibilities for protein crystallography i.e. as reported in Andrews *et al.* (1988). Subsequently Ohsumi *et al.* (1991) designed a diffractometer working in vacuum for submicrometre chemical crystallography.

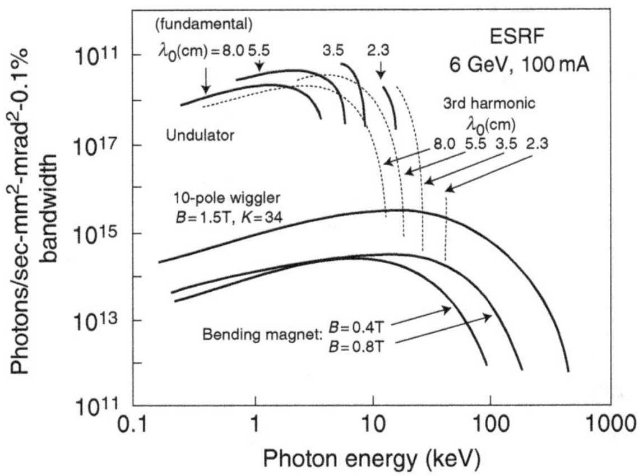

Fig. 16.5. Brilliance of the early ESRF sources as a function of energy (From Altarelli (1991) with permission of the author and Physica Scripta).

Helliwell (1992) (pages 412 to 413) tabulates the sample scattering efficiencies comparing small molecules, proteins and virus small-crystal cases and resulting very weak scattering.

The 3rd-generation source, undulator, monochromatic X-ray intensities that one was considering in this context in the mid-1980s planning discussions were so large (Figure 16.5) i.e. in the range of 10^{14} or more onto the sample, that this caused anxieties as to the radiation and thermal damage that might be expected in such beams hitting biological crystallography microcrystals. A convenient way to assess these possibilities at the time, and thereby assist the forward planning, was to use Laue diffraction from 2nd-generation wiggler source polychromatic beams, to mimic planned monochromatic beam intensities, such as from the SRS wiggler 9.6 station. Thus, tests were made with a 20-micrometre on edge gramicidin test crystal (Hedman *et al.*, 1985); this study showed that large swathes of reciprocal space could in fact happily be sampled without undue damage problems for the sample. The ESRF Foundation Phase Report (1987) plan then (see, e.g. Altarelli (1991)) for the macro-molecular crystallography beamlines provision, included a facility for measuring from microcrystals. Other studies very usefully extended the detailed considerations of how small a protein crystal one could use for data collection in ever more intense monochromatic beams (Glaeser *et al.*, 2000). Also, in the more specialized application of freeze trapping, as Teng and Moffat (1998) pointed out, '*The smaller the sample size, the lower the thermal diffusion time and the more rapid the cooling*' thus allowing more prompt reactions to be investigated by that time-resolved technique.

Thus, as Altarelli (1991) reported 'on a microfocus ESRF beamline focal spot sizes of 10×10 micrometre, and via further collimation using slits, beams of 1×1 micrometre are envisaged for microcrystallography, small-angle scattering and microtomography and with photon energies up to 25 keV'. A microfocus beamline

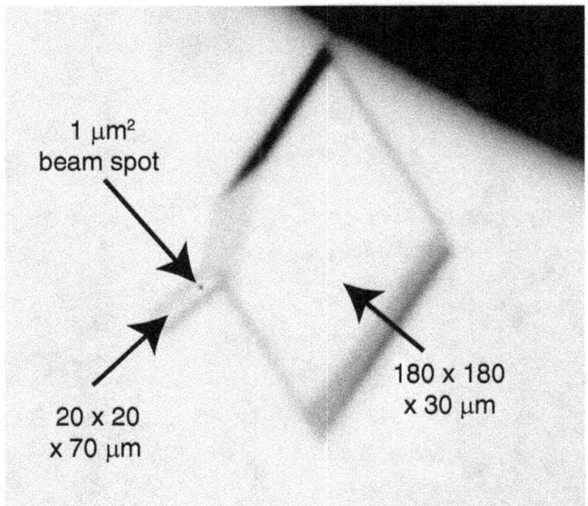

Fig. 16.6. A test case for data collection involving a 1-micrometre X-ray beam spot (from Moukhametzianov *et al.* (2008)). With permission of the authors and the IUCr.

was indeed developed by Riekel and coworkers at ESRF and found numerous applications (e.g. Figure 16.6 used from Moukhametzianov *et al.* (2008)).

A specially developed microdiffractometer, to control the background scatter and achieve much more precise crystal sample manipulation than hitherto, was developed by EMBL with ESRF, Figure 16.7, (Perrakis *et al.*, 1999).

This diffractometer became commercially available and has been installed on various 3rd-generation SR sources for example on the APS and DLS. Further important optimizations on collimation selection have then been introduced by Fischetti *et al.* (2009) at GMCA CAT on APS (Figure 16.8).

A further remarkable development is the screening for good diffraction of microcrystals sitting directly in a beamline-mountable crystallization well (Lunde *et al.*, 2008)

Microcrystal experiments require robust beam stability during data collection and minor fluctuations cannot be ignored and effective countermeasures need to be taken (see, e.g. Igarashi *et al.* (2008)).

In estimating how far one can go in reducing the sample size, there are several considerations including sample manipulation, imaging and radiation damage in the beam. Optical tweezers (Singer *et al.*, 2004) are being investigated for manipulation and for visualization the visible light wavelength becomes important at very small sample sizes. Microscopes are being pushed towards the short-wavelength end of the visible spectrum and even the UV (Lunde *et al.*, 2005). It is also easier to see if something moved upon deliberate translation in an image rather than a static 'dot' (Riekel, personal communication).

In terms of radiation damage a thought-provoking analysis has been provided by Nave and Hill (2005), based on Henderson's observation of the range of photoelectrons in a protein sample (Henderson, 1990). A sample may be small enough for the damaging photoelectrons to immediately escape the sample! Nave and Hill's

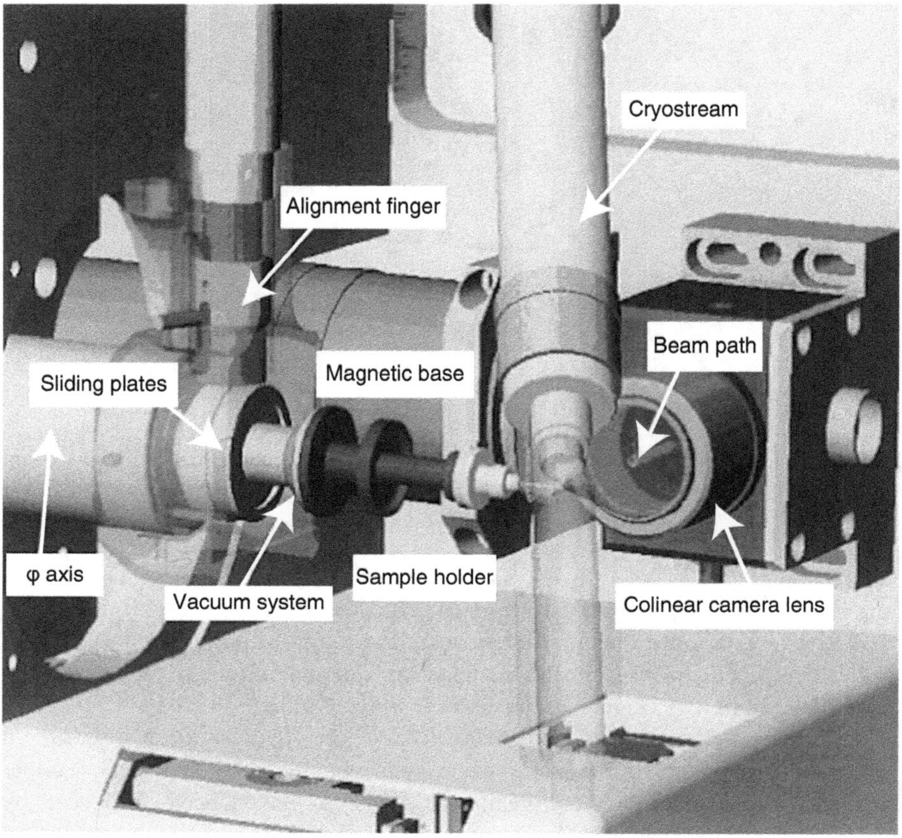

Fig. 16.7. A schematic of the ESRF EMBL Microdiffractometer (from Perrakis *et al.*, (1999) used with the permission of the authors and the IUCr.

calculations indicate that significant reductions in X-radiation damage could occur for crystals less than a few micrometres in size. The benefits would be greater when operating at higher X-ray energies (Table 16.1).

Other methods approaches are possible. There may yet be a resurgence of electron crystallography in this field of microcrystals (Hovmoller, personal communication; Abrahams, personal communication) since the scattering interaction of a sample is strongest with electrons. Thus, as studied by Dorset (1975) originally, Dimmeler *et al.,* (2000) for example showed electron crystallography measured experimental data of frozen hydrated crystals of the protein catalase. Perhaps most exciting is to consider on an SR microfocus and submicrometre focus beamline the capability of being able to switch to the option of an electron microbeam on the X-ray beam axis. A combination of electrons and X-rays for microcrystallography would allow the gathering, comparison and usage in complementary ways of the necessary diffraction data sets on this challenging and yet important samples' frontier.

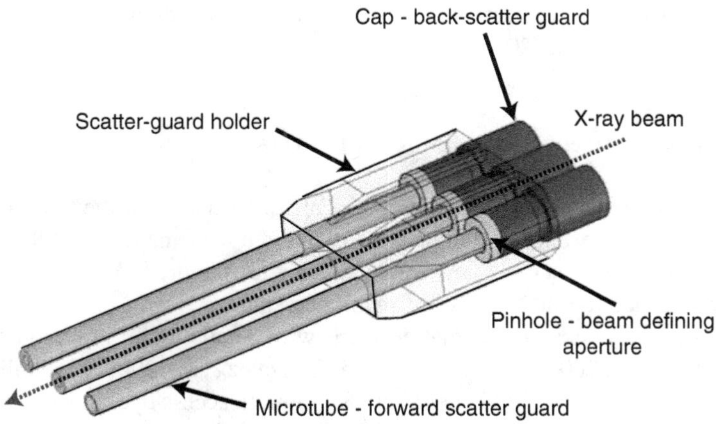

Fig. 16.8. The triple collimator installed at APS GMCA CAT (Fischetti *et al.*, 2009). Used with permission of the authors and the IUCr.

16.5 Phasing and the Available Instrumentation

The fundamentals of phasing are covered in various textbooks that are available. A review of the topic can be found in Taylor (2003). A historical overview of beamline provision, notably including rapid tunability at increasingly high X-ray brilliances for MAD phasing is given in Cassetta *et al.*, (1999). A particularly flexible recent beamline was the SRS Beamline 10 (Cianci *et al.*, 2005a) and that included tilted detector geometry for longer-wavelength work including sulphur SAD phasing (Cianci *et al.*, 2008). For conventional-source apparatus the reader is referred to the various manufacturers websites and leaflets. An extensive survey of applications of

Table 16.1. The percentage of energy that escapes from a sample as a function of the initial energy of the photoelectron and its initial distance from the surface of the protein microcrystal (From Nave and Hill (2005) with permission of the authors and the IUCr).

Distance from surface (um)	Transmitted energy (%) at given energy			
	10 keV	20 keV	30 keV	40 keV
0.5	79	95		
1	23	89		
2	0	74	90	94
4		34	76	87
8		0	37	57
16			0	23
32				0

resonant scattering in structural chemistry and biology including phase determination is given in Cianci *et al.*, (2005b).

16.6 Robotics, Telepresence and Remote Access

A recent development in synchrotron data collection has been the use of robotics coupled with telepresence techniques. This has reduced manual handling significantly and increased sample throughput such that hundreds of samples can be routinely screened to find the best-diffracting crystal.

The most mature example is that at SSRL with details found in various publications (Cohen *et al.*, 2002; Cohen *et al.*, 2005; Gonzalez *et al.*, 2006; McPhillips *et al.*, 2002; Miller *et al.*, 2004; Soltis *et al.*, 2008). Remote access operations at SSRL for the general biological community started in 2004. This has been successful with over 75% of users now collecting data remotely from more than 200 000 crystals to the time of writing this book. The approach taken is remote access rather than telepresence (defined below). Cassettes or pucks of samples are loaded into the beamline dewar by the synchrotron staff, control is handed to the users working from a remote site, and then data collection is completely controlled by the users with intervention only in the case of technical difficulties. To achieve this significant efforts have been put into control software, video monitoring, robotic hardware and data-collection software. Significant effort is also involved with the beamline staff in maintaining the instrumentation and regular beamline diagnostic work to ensure that users have no issues that would delay or halt data collection. Similar efforts are underway at other synchrotron sources, e.g. (Beteva *et al.*, 2006; Cork *et al.*, 2006).

Another approach is telepresence, the historical development of which is described briefly in Warren *et al.* (2008) including various citations to microscopy and chemical crystallography, as well as macromolecular crystallography, developments since 1992. In their system Warren *et al.* incorporated a dedicated communication system at the SRS chemical crystallography station 9.8 that included visual *and* audio systems for those present at the beamline linking them to others not at the site. This has considerable advantages in that expertise can rapidly be sought from team members not present that might be especially familiar with a particular and perhaps unexpected aspect of the experiment. It was also shown to have a particular use in training users for synchrotron operation. By contrast, the technologies developed at SSRL make use of chat rooms for similar communication but video output is only from the beamline to other users and not vice versa. In the SSRL case of remote access rather than telepresence typical full remote data collection does not involve any presence of the experimenter at the beamline.

Other programs related to this have been termed Mail-in or FedEx crystallography. Key to their success is the ability of the experimenter to monitor the diffraction results, and either make decisions on the next step, or offer input on those decisions. Again a number of synchrotrons have been developing these monitoring tools (Okazaki *et al.*, 2008; Skinner *et al.*, 2006; Yamada *et al.*, 2008).

From a practical point of view, these technologies, while increasing throughput and minimizing handling, also allow more thought to be given to the experiment. The experimenter performs the experiment from the home laboratory where they have immediate access to all the details of the experiment, collaborators involved can add their input after observing the data, and all this can be achieved in a more relaxed atmosphere than at the beamline.

16.7 More Specialized Applications

More specialized applications in macromolecular crystallography data collection require consultation of more specialized reviews, although similar 'measurement physics' principles apply. Special features are briefly indicated in the list below for each case:

1. Neutron beams involving steady-state reactor Laue (Blakeley *et al.*, 2004) or time-of-flight Laue (Langan and Greene, 2004; Langan *et al.*, 2004) or steady-state reactor monochromatic (Niimura and Bau, 2008) are used for hydrogen and hydrogenation (protonation) studies of biological macromolecules; special features include growing larger crystals than for X-ray work as well as either D_2O soaking or microbiological production of fully deuterated proteins and use of fully deuterated reagents; in the Laue case special software packages have been developed for wavelength normalization for synchrotron X-ray Laue diffraction (Helliwell *et al.*, 1989) and were readily ported to analyse neutron Laue data (changes to the polarization correction are needed for neutron versus synchrotron X-ray beams and at neutron facilities cylindrical image-plate detector geometry is preferred that obviously requires changes of the diffraction spot coordinate predictions). Notably with neutrons there is no sample absorption or radiation damage.
2. Time-resolved X-ray Laue diffraction uses short pulses of X-rays either from synchrotrons (100 ps) (Ren *et al.*, 1999) or Compton back-scattering sources (100 to 300 fs) (Priebe *et al.*, 2008); special software packages have been developed for wavelength normalization and harmonic reflection intensity extraction (see Ren *et al.* (1999) for a review); spot sizes can show time-dependent changes, e.g. as acoustic shock waves traverse a crystal (Edwards *et al.*, 1990); the emergence of CCDs in the 1990s radically altered measuring capabilities here in both reducing the radiation dose to the sample for a desired intensity precision in the diffraction spots as well as allowing full automation of the many cycles of 'pump and probe' style of experiment (see, e.g. Nieh *et al.*, (1999)).
3. Freeze trapping of structural states, i.e. including intermediates is made via combined UV/vis spectroscopy and crystallography (e.g. Bourgeois and Weik (2009)); spot sizes can increase upon freezing samples as well as increasing rocking curve widths.
4. X-ray free-electron lasers are becoming available to provide 10-fs pulses (Neutze *et al.*, 2000); see Chapter 19.

5. Electron crystallography (see, e.g. Dorset and Parsons, 1975; Glaeser *et al.*, 2007); in recent years there has been the introduction of the precession method in electron crystallography, and that is leading to significant improvements in the quality of the intensity measurements not least in making the electron scattering from the crystal much more kinematical; there is important potential here in measuring data from microcrystals, even sub-micrometre crystals, as mentioned in Section 16.5.

16.8 Chapter Summary

This chapter has had to be more indicative than detailed as the topic is large. The consultation of the example references given and wider literature is obviously important. Nevertheless, in the context of macromolecular crystallization and crystal perfection the chapter connects directly with the data-collection requirements for structure determination and offers useful pointers in that direction in the various categories.

17

Protein powders – making the most of tiny crystallites in bulk

17.1 Introduction

Crystallization often produces microcrystals where single crystals suitable for X-ray diffraction studies are not obtained. An assembly of these microcrystals produces a crystalline powder. Crystalline powders diffract and typically produce a series of rings called a powder pattern. There is the potential to make use of this powder data where it proves impossible to coax the conditions into those suitable to produce single crystals; a notable case considered by Plohn and Buldt (1986) involved membrane proteins. Various advances have been made in the field of structure solution from powder diffraction data notably through the work of von Dreele and then at ESRF by Margiolaki and Wright on beamline ID31. These relatively recent developments are in the context of significant advances in the last two decades. Instrumentation for the use of finely collimated synchrotron radiation yields better resolved powder lines that have been found to be beneficial in the structure solution of smaller molecules (less than $\sim$30 non-hydrogen atoms), for example, cimetidine (Cernik *et al.*, 1991). The use of variable-temperature measurements also proved useful exploiting anisotropic thermal expansion, and thus reduced peak to peak overlap problems (Shankland *et al.*, 1997b). There have in addition been advances in structure solution methodology with powder data e.g. use of Monte-Carlo methods (Harris *et al.*, 1994), simulated annealing (David *et al.*, 1998; Pagola and Stephens, 2000) and genetic algorithms (Kariuki *et al.*, 1997; Shankland *et al.*, 1997a). Despite these advances, *ab initio* structure solution from powder diffraction data is still not routine in small-molecule studies, particularly for more complex structures; this is because, generally, there are a significant number of overlapping reflections, in addition to the fact that Friedel pairs of reflections are exactly coincident. Thus, there tends to be an underdetermination of data, where the data tends to be weak, and therefore does not routinely extend to atomic resolution. It is remarkable then that protein powder diffraction of crystal structures with $\sim$1000 non-hydrogen atoms in the asymmetric unit are now being tackled.

Margiolaki and Wright (2008) review the topic, including a historical survey of examples of protein powder patterns stretching back some 60 years; an example is that of Amos *et al.,* (1984) who showed the powder patterns, measured with a conventional X-ray source, from tubulin microcrystals, Figure 17.1.

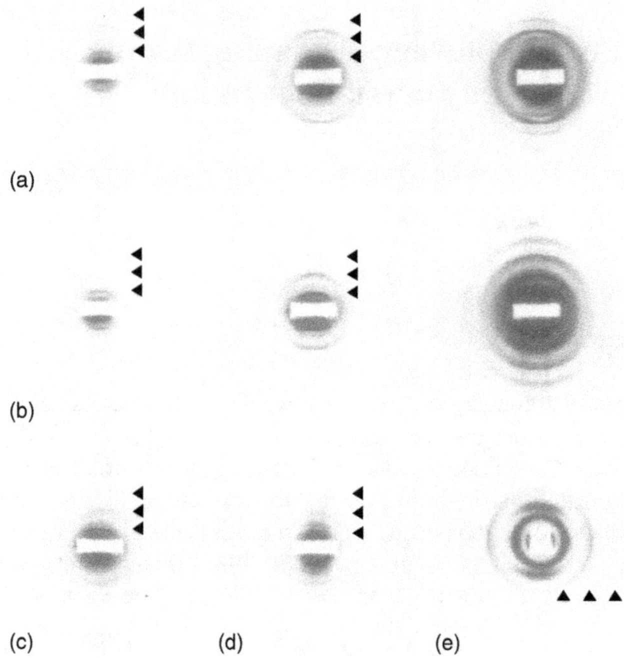

Fig. 17.1. Typical X-ray powder patterns of tubulin, pelleted, crystals. Arrowheads point to the positions of the (1 0 0) (2 0 0) and (3 0 0) reflections. For patterns (a) to (d) the X-ray optics comprised a mirror and a quartz monochromator to give very good vertical resolution. For (e) a pair of mirrors gave isometric resolution though less sharp than in (a) to (d). Powder patterns from crystals of brain protein tubulin, though closely related to those of sea urchin egg crystals, were shown to be distinctly different (data not detailed here). All but one of the strong and many of the weaker reflections indexed on a hexagonal lattice with a unit-cell spacing of 373 (+ 3) Å. The remaining strong reflection and most of the other weak reflections were consistent with a c-axis spacing of 144 (+ 4) Å. From Amos *et al.*, (1984) with the permission of the authors and the *J Molec Biol.*

17.2 Quantitative Protein Powder Analyses have Opened up in Recent Years

The more recent times have moved from unit-cell parameter determination and relatively qualitative interpretation of protein powder patterns, although clearly valuable, to detailed protein powder profile intensity and reflection intensity extraction via deconvolution analysis. Thus in the protein powder work of von Dreele, including the important development of the program 'GSAS', applied to proteins such as insulin and hen egg white lysozyme, first, it was shown that protein model refinement against protein powder data was possible up to approx 3 Å resolution (Von Dreele *et al.*, 2000) and then that ligand difference electron density could be resolved. Margiolaki *et al.,* (2005) showed that molecular replacement structure solution was possible with protein powder data again with lyzozyme. Figure 17.2 shows an image-plate experimental setup and measured lysozyme powder pattern and derived intensity profile versus 2θ (Von Dreele *et al.*, 2006).

Most recently a 67 amino acid residue protein, ponsin, has been solved from protein powder data (Margiolaki *et al.*, 2007b), where the extracted intensities were of sufficient accuracy to employ established single-crystal molecular replacement methods with a search model with about 40% amino acid sequence identity.

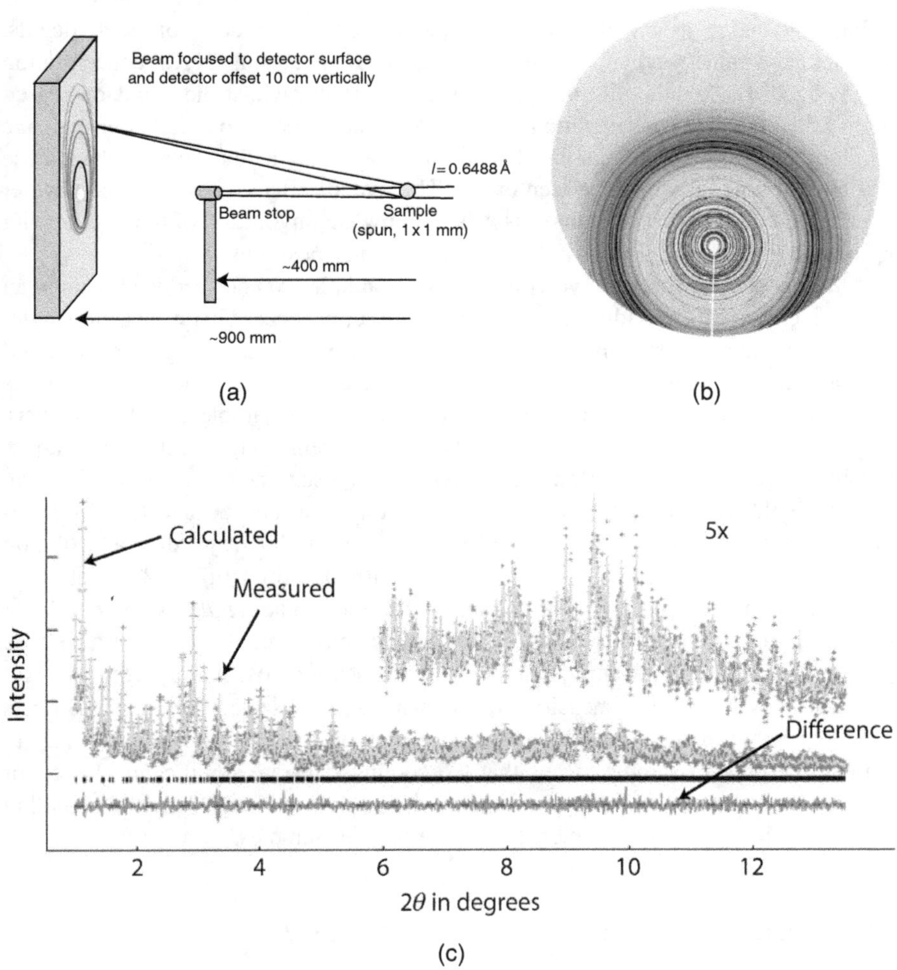

Fig. 17.2. (a) Schematic of powder diffraction setup adapted from von Dreele *et al.* (2006) with (b) data from lysozyme (Margiolaki *et al.*, 2007a) and (c) an example pattern with experimental points as crosses and a Rietveld fit as the continuous line – the data has been enlarged 5× in the top right. Used with the permission of the authors and *Z. Kristallogr.*

The widening of the applicability of the use of protein powder data for structure determination is envisaged notably via the use of multiple powder patterns under variable conditions, e.g. of temperature. von Dreele (2007) showed that by using combinations of multiple protein powder diffraction patterns, which differed by solvent-induced and radiation-damage-induced lattice strains, the severe protein powder overlap problem up to a d-spacing of 2 Å was partially resolved in a stereochemically restrained Rietveld refinement for chicken egg white lysozyme including yielding placement of a substantial number of bound water molecules, a remarkable achievement. This study also explored the lattice strains induced by changes in salt concentration, changes in solvent pH and the effect of low-dose radiation damage as the powder technique is very informative on such details. The lattice strains were not monotonic, i.e. with increasing NaCl concentration (0.25–1.25 M) the unit cell *a*-axis increased by ~0.5%, whilst the *c*-axis decreased by ~1.5%, and also the variation was pH dependent. Low-dose radiation damage similarly induced non-monotonic lattice strains, similar to but smaller than those arising from increased salt concentration. The overall effect of using these powder data in a combined Rietveld analysis was to allow deconvolution of the overlapping reflections by differing shifts in their relative angular positions.

A further step is whether cryo-cooled powder samples can be produced without an undue increase of the width of the powder diffraction lines. Given that the single-crystal experience is that smaller crystals freeze more easily than larger crystals, and with a smaller increase of intrinsic mosaicity, the case for the powder situation looks promising. However the crystallites are gathered in a bulk sample, i.e. akin to a very large single crystal. Nevertheless, the first experiments have been encouraging, notably by Jenner *et al.*, (2007); the survival of the lysozyme protein powder test in the ESRF ID 31 beam was increased by a factor of around 30 and thereby high-quality powder diffraction patterns could be collected at 100 K, yielding a resolution of $d_{\min} \approx 2.6$ Å, significantly better than the previous limit of ~3.27 Å at room temperature, and despite a smaller volume of sample. Jenner *et al.* also reported that the powder diagrams were well accounted for by considering only a 'strain'-related peak broadening term, meaning that the peak broadening arose mainly from having a distribution of unit cell dimensions in the sample. Particle-size broadening effects were negligible in comparison. The increased sample size and use of a capillary to contain the powder sample means that the cooling rates are significantly lower for powders in comparison to single crystals in loops and challenges remain in finding ways to optimally cryo-cool microcrystalline powder samples.

17.3 Structure Determination of Unknown Proteins?

The solution *de novo* of novel protein structures, anticipated by Plohn and Buldt (1986) in their consideration of anomalous scattering, can also be considered via isomorphous replacement. Thus, von Dreele *et al.* (2006) have investigated xenon binding to lysozyme using powder data. Protein molecular envelopes have thus been obtained at 6 Å resolution using a uranyl derivative of elastase (Wright *et al.*, 2008). Helliwell *et al.* (2005) have considered the case in detail of powder dispersive differences (PDD) in the context of the multiple-wavelength anomalous dispersion (MAD) single-crystal method, now routinely employed for the single-crystal structure solution of macromolecules. MAD exploits the changes in the atomic scattering factor that take place close to the absorption edges of elements, due to the variation of f' and f'' with wavelength, where the atomic scattering factor is given by:

$$f = fo + f'(\lambda) + f''(\lambda) \tag{17.1}$$

and f_0 is the basic scattering factor of an atom and $f'(\lambda)$ and $f''(\lambda)$ are its real and imaginary components of the anomalous scattering. By tuning to the absorption edge of a specific element in a sample to stimulate the maximum changes in f', it is possible to induce differences in reflection intensities with wavelength, similar to those seen with the isomorphous replacement method (see, e.g. Helliwell (1992) and Ramakrishnan and Biou (1997) for such a treatment). Approximately a third of all genes are thought to code for metalloproteins, thus providing an anomalous scatterer in these cases. On the theoretical basis it must be stressed that a fundamental limitation of the powder method remains even with a two-wavelength approach, since the hand of a molecule cannot be determined because the Friedel equivalent reflections exactly coincide in a powder diffraction ring. In the protein case, it is known from the chemistry that in nature, amino acids are left-handed, this chemical property will be valuable in structure determination re choice of the hand. Nevertheless, additional sources of phase information are needed. The two, or more wavelengths, $\Delta f'$ induced reflection intensity changes will yield a phase ambiguity. Wang (1985) considered the case of applying solvent flattening to extract the correct protein electron-density image in the case where there is such a phase ambiguity per reflection in single isormorphous replacement (SIR) and that is formally equivalent to the change in f' case.

17.4 Characterization of Protein Crystal Polymorphism

Impetus in this aspect has so far been from the pharmaceutical industry. Such polymorphs form the basis for patent applications as the powder slurry is closest to the physical state of the sample used by patients. Thus, Norrman *et al.* (2006) have used a laboratory, i.e. CuKα X-ray powder diffractometer to characterize twelve different microcrystalline insulin formulations by X-ray powder diffraction. Three of the insulin formulations crystallized in the same crystal system, but had structural differences in the N-terminal B-chain of the insulin molecule. This difference was efficiently detected in the powder patterns. The sensitivity of the method made it a valuable tool for characterization of the microcrystalline samples. By use of principal-component analysis, the twelve different insulin formulations, originating from six different crystal systems, were classified into nine separate analysis clusters.

The powder patterns of each analysis cluster could be used as 'fingerprints' for the different insulin crystalline polymorphs. The combination of X-ray powder diffraction and multivariate analysis, such as principal-component analysis, provided a rapid and effective tool for studying the influence of chemical derivatives, additives, ions, pH, etc., in the crystallization media.

17.5 Chapter Summary

As Margiolaki *et al.* (2007a) state "Determination of protein crystal structures is dependent on the growth of high-quality single crystals, a process that is not always successful. Optimum crystallization conditions must be systematically sought for, and microcrystalline powders are frequently obtained in failed attempts to grow the desired crystal. In materials science, structures of samples ranging from ceramics, pharmaceuticals, zeolites, etc., can nowadays be solved, almost routinely, from powdered samples, and there seems to be no fundamental reason, except the shear size and complexity of the structures involved, why powder diffraction should not be employed to solve structures of small proteins." Obviously reflection intensities have to be extracted for structure determination. Multiple pattern analyses have been employed by von Dreele (2007) and the cryo-cooling of powders is a significant development made by Jenner *et al.* (2007). Both of these developments will be important indeed vital components for the applicability of isomorphous replacement or of use of dispersive anomalous differences. The tubulin example quoted in the opening of this section is of a very important protein, involved in cilia and flagella motion, for example, and it took structure solution using single-crystal techniques many years. One wonders what if such microcrystals could have enjoyed the access then to the 3rd-generation X-ray source beamlines. It is an open question whether micrometre sized single-crystal techniques would be the better way forward or the powder methods. Pros and cons would probably be in favour of robotic control of placing many individual microcrystals into a micrometre-sized beam ie one by one. Nevertheless the powder method is a challenge that is being taken up. The pharmaceutical companies (see, e.g. Norrman *et al.,* (2006)) are notable as players in this unfolding story of new methods applied to old and new challenges.

How common are microcrystals? To date, there has been no systematic study published of macromolecular crystal size over a wide range of different sample types. However, Richard Gillian (MacCHESS) is in the process of examining crystal-size distributions from high-throughput crystallization screening trials of 68 different macromolecules and over 100 000 photomicrographs of crystallization conditions (68 proteins, each in 1536 different potential crystallization conditions). For this particular batch mode of crystallization (Luft *et al.*, 2003), the finding was that "the largest crystals for each protein are comparable to $50\times 50\times 50$ micrometre^3 of illuminated volume or less and comprise 55% of the samples. The frequency of small crystals appears to increase exponentially with decreasing size within the limits of the statistics." Thus, microcrystallography looks set to be a very significant growth area in the field. Working with such crystals may be envisaged as one by one, several in the X-ray beam at once or use of them in a bulk powder. Combinations of such techniques may also be appropriate. The detailed case of microcrystallography and the appropriate instrumentation for X-ray data collection from such single-crystal samples are discussed in Chapter 16.

18

Complementary techniques

In this book we have discussed crystallography in great detail. It provides a powerful window into the biological world at an atomic level with over 60 000 structures determined to date. However, where both failure and success is tracked in structural genomics efforts, currently less than 15% of purified targets result in crystal structures. Similarly we should also remember that a crystallographic structure is a time and spatially averaged model of the actual molecular structure and it is derived from a state where crystallization contacts can affect the interpretation of the structure.

Fortunately there are a number of complementary techniques that can help us in determining structures where we do not have crystals or where we would like to build on or validate the crystallographic information. Some of these are nuclear magnetic resonance (NMR), X-ray absorption spectroscopy (XAS and EXAFS) and small-angle X-ray scattering (SAXS). NMR provides atomic level detail for smaller macromolecules, SAXS provides low resolution macromolecular envelopes but applicable to all samples.

NMR is primarily a solution technique. In the simplest terms NMR measures the absorption of discrete frequencies as an atom resonates. Individually identical atoms would resonate at the same frequency but in an assembly, i.e. the macromolecule, each atom is shielded by the surrounding electrons from others. This electronic shielding reduces the resonate frequency and provides a direct probe of the chemical environment. This is termed the chemical shift. NMR techniques applied to proteins are covered in detail in many publications, e.g. Cavanagh *et al.* (2006). NMR offers protonation state titration curves to determine the pKa values of ionizable amino acids (His, Glu, Asp, Arg and Lys). NMR also allows ligand binding to proteins to be measured even at weak affinity.

The complementary nature of NMR to crystallography has been studied with a large number of samples. Yee *et al.,* (2005) determined that NMR was effective as a screening technique to determine if the sample was well folded, if it formed higher-order oligomers, non-specific aggregates or conformational fluctuations on the NMR timescale. Snyder *et al.* (2005) compared the correlation between NMR data quality and success in structural determination by crystallography. They found that NMR data quality did not correlate with crystallization success. NMR provided a completely complementary technique to extract structural information where crystallization failed. Conventional NMR is limited to smaller macromolecules, up to ~100 kDa although specialized methods are being developed to extend this limit further (Tzakos *et al.*, 2006).

X-ray absorption spectroscopy (XAS and EXAFS) gives precise local distances around metal centres. The dose required is less than in X-ray crystallography and has found notable application to characterizing radiation damage in metal sites.

SAXS is a low-resolution technique and is limited in that multiple models can be fitted to a single scattering curve. Its power comes where additional information is available. The information SAXS can provide and the information in combination with crystallography is reviewed in detail by Putman *et al.,* (2007). SAXS combined with high-resolution crystal structures, establishes architectures, assemblies, conformations, and unstructured regions for proteins and protein complexes, e.g. Tsutakawa *et al.,* (2007). SAXS relies on the interference between X-rays elastically scattered between spatially correlated electron pairs. As a solution technique, molecules are randomly arranged in solution and are in motion. SAXS data is averaged over this and is symmetric around the incident beam. It is an averaged structure but sensitive to dynamical information. Neutrons can also be used for small-angle scattering. In some cases, the combination of techniques, SANS, SAXS and X-ray crystallography is the only method of obtaining pertinent structural information (Forstner, 2000).

Other techniques that can help the crystallographer include cryo-electron microscopy. Navaza (2008) showed how cryo-electron microscopy data can be used as a search model to resolve the phase problem in molecular replacement techniques and as a means to build up multimeric structures from individual components.

18.1 Chapter Summary

Crystallography is a powerful technique but also one that has limitations. There are a number of complementary techniques that significantly add to the information available crystallographically.

PART IV

THE FUTURE

19

The X-ray laser and the single molecule – no crystal needed?

Major increases in X-ray brilliance, compared with the most brilliant 3rd-generation SR sources, are anticipated with the next generation of X-ray lasers (Figure 19.1). The gain factor in effect, i.e. by a simple extrapolation, means that the sample size can reach the size of a single molecule that would generate measurable scattered X-ray diffraction pattern intensity. The diffraction pattern also is governed by the continuous Fourier transform of the diffracting molecule. As suggested by Sayre (1952), by giving access to 'half-order' intensities, this 'oversampled' continuous diffuse scattering allows solution of the phase problem, as first demonstrated experimentally by Miao *et al.* (1999). The resulting method of 'lensless imaging' for non-periodic objects has since achieved 10 nm resolution in three-dimensional (tomographic) image reconstructions using soft X-rays (Chapman *et al.* 2006). The iterative phasing methods used (which have much in common with electron density modification) are based on variants of the Fienup method (Fienup (1982)) (See Spence (2007) for a review).

These phasing methods have also been used to phase the continuous diffraction patterns produced from a non-periodic sample by a single femtosecond X-ray pulse. As suggested by Neutze *et al.*, (2000), a sufficiently brief and intense X-ray pulse will terminate before radiation damage commences, yet provide sufficient photons for a useful pattern. The duration must be shorter than the timescale of nuclear motion (the Debye period of about 100 fs) and the time taken for significant redistribution of the electron density (often taken to be the Auger recombination time of a few fs). There is now clear evidence in support of this 'diffract-and-destroy' principle, at least at low resolution (Chapman *et al.* (2006), and the first image of a biological cell using this method was recently published (Marchesini *et al.* (2008)). Since the sample is subsequently destroyed, a series of identical particles is needed in different orientations. The problem of determining these orientations from their diffraction patterns has much in common with a similar solved problem (with images, not diffraction patterns) in single-particle cryo-electron microscopy. The problem, for X-ray diffraction, has recently received considerable attention (Schmidt *et al.*, 2008; Fung *et al.*, 2008). Experimentally, a single-file beam of hydrated molecules is passed across the X-ray laser, possibly in synchrony with it, and a single diffraction pattern is read out in milliseconds after each pulse (Bogan *et al.* 2008). Details of one experimental

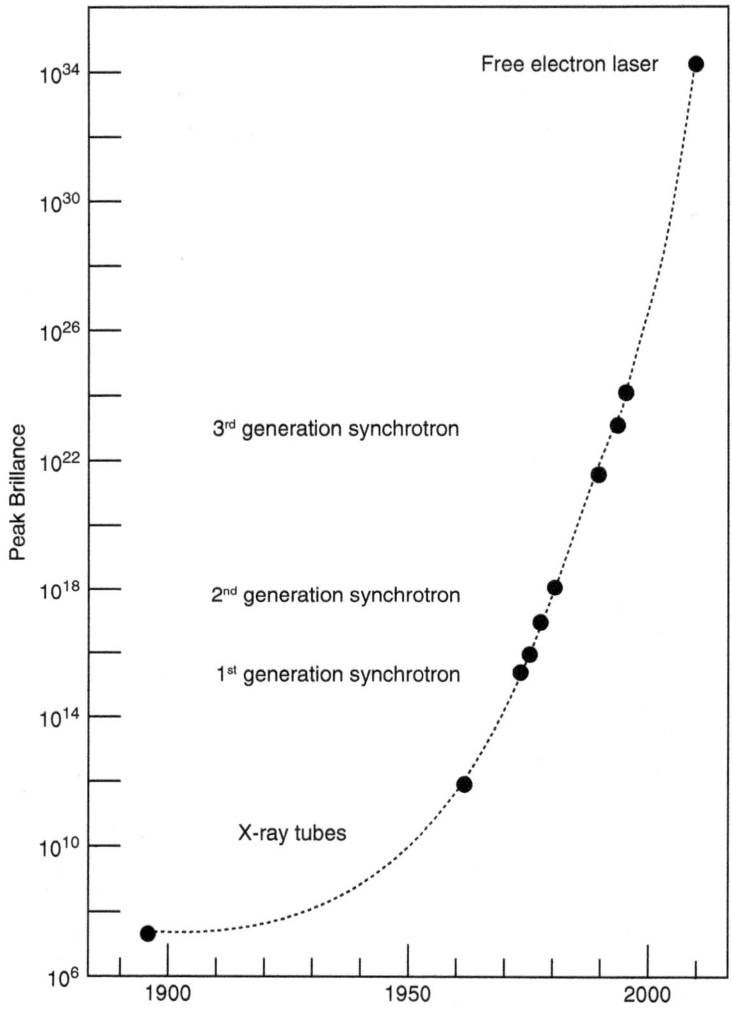

Fig. 19.1. X-ray brilliance of example X-ray sources versus calendar year.

hydrated protein beam generator have been given (Shapiro *et al.* 2008). This 'aerojet', using gas focusing and driven by a piezo vibrator, will be used at the Linac Coherent Light Source (LCLS, the first hard X-ray FEL). Alternatively, a continuous X-ray beam might be used with this single-file molecular beam, while laser-alignment is used to align the molecules. This has been addressed in detail by Spence and coworkers (2005) who advocate use of a medium-energy SR undulator, with a molecular beam of carefully controlled aligned molecules using a polarized laser beam (Figure 19.2). Thus, a stream of molecules suitably oriented pass an electron or

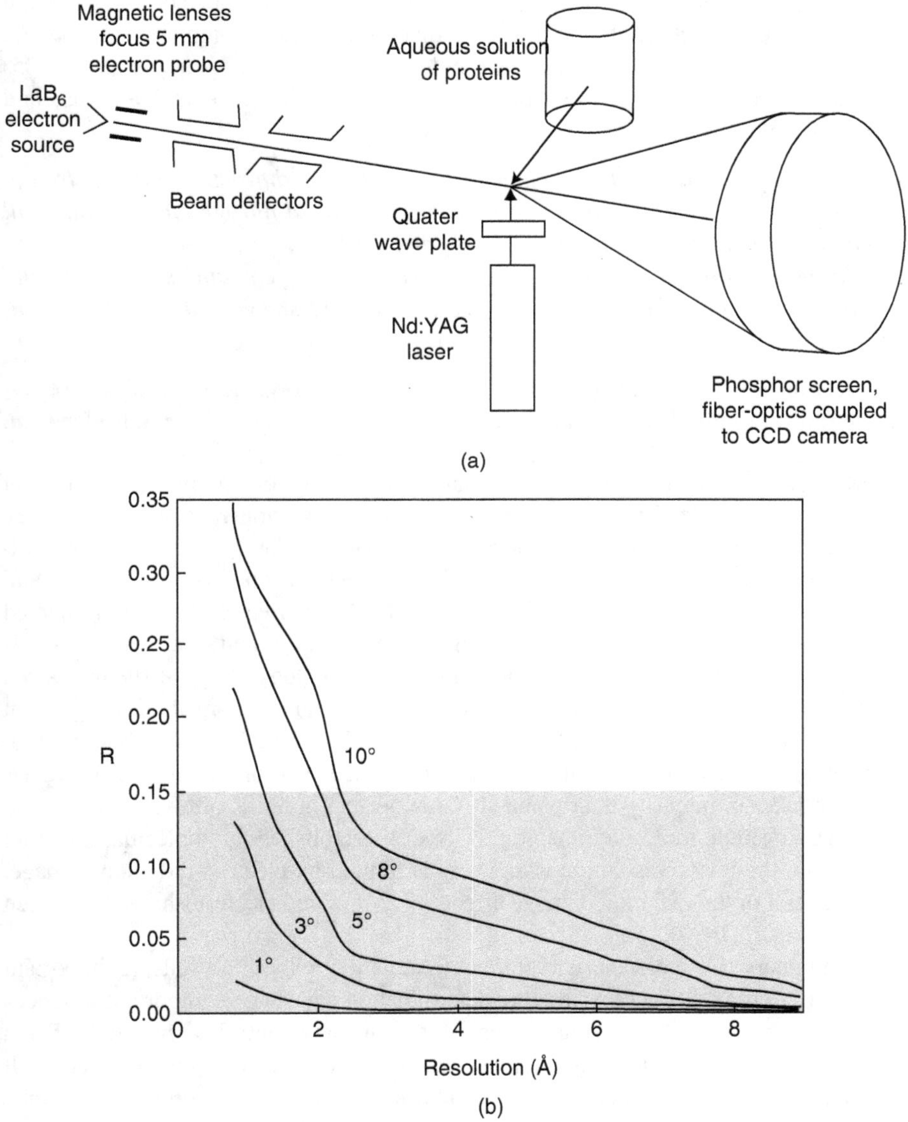

Fig. 19.2. Spence *et al.*, (2005) advocate a molecular-beam arrangement shown in (a). Hydrated organic molecules emerge from a nozzle into vacuum where droplets form and cool by evaporation to form a beam of doped ice balls traversing the electron beam. A continuous-wave fiber laser is used to align them. The laser polarization is rotated after each data readout for a new molecular-beam orientation. They have also addressed the question of misorientation errors of molecules in a molecular beam in terms of R_{factor} increases for a test molecule of CuPhTh. There is a strong dependence of R_{factor} degradation with resolution. Spence *et al.* consider $\boldsymbol{R} = 0.15$ to be the minimum acceptable value for a faithful reconstruction. From Spence *et al.*, (2005) and used with the permission of the authors and the IUCr.

X-ray beam and diffraction data are accumulated until sufficient statistical precision for that pattern is achieved, and thereby radiation damage is completely avoided. A new pattern is then acquired at the next molecular orientation. Spence *et al.'s* (2005) simulations include small molecules, a protein, the ribosome and a virus, and conclude as follows:-

"In summary, we find that sub-nanometre resolution in diffraction patterns from a laser-aligned molecular beam of small proteins at liquid-nitrogen temperature will require kilowatt CW laser power in the infrared, whereas tens of Watts are needed for large macromolecular assemblies and viruses. The conditions of power and temperature depend strongly on the molecular size, shape and (poorly known) dielectric constant, but most importantly on size."

Electron diffraction from a stream of laser-aligned organic molecules (not hydrated) has recently been demonstrated using a pulsed system (Reckenthaeler *et al.* (2009)).

Miao *et al.* (2001) outline practical details of using the USA X-ray Laser Coherent Light Source at Stanford, which with its planned repetition rate of 120 Hz, they estimated a data-acquisition time of diffraction patterns for 10^6 copies of the molecules at about 2.3 h. This is not a long data-acquisition time, to say the least, and thus it would be a powerful combination of the X-ray free-electron laser-enabled diffraction imaging and the oversampling phasing method and that therefore obviously could have an important impact on structural biology. They also emphasized that their approach is a "general 3D structural determination method and can, in principle, be applicable to electron and neutron diffraction for imaging both biomolecules and nanoscale materials". Currently planned neutron beams are, however, much weaker in intensity than planned X-ray beams. In terms of electrons multiple orientation single-molecule imaging is a well-established biomolecular imaging technique; the image resolution achievable is limited by electron radiation damage. Comparison of the radiation damage effects of X-rays and electrons have been given by Henderson (1990).

In summary, for molecular structure determination without crystals, an important change is afoot in diffraction structure determination ideas where important advances are taking place in X-ray source technology, in laser-controlled molecular-beam alignment and theory and simulations, where the crystal may not be needed. It remains to be seen in practice what structure determination resolution is achievable, and specifically how the new methods will compare with existing electron microscopy techniques using cryo-preservation of hydrated biological samples.

The above advances are also relevant to time-resolved structure determination. As for molecular structure determination without a crystal two (at least!) conceptions are possible. First, the X-ray laser exposure time per diffraction image is clearly in the fs range. Thus, by appropriate time lag between a fs laser pumping pulse and the probing X-ray fs laser pulse fs time-resolved structure determination might be possible. Secondly, with the laser-controlled aligned molecules in a molecular beam a second, e.g. fs laser could pump the protein and after an appropriate time lag the X-ray (or electron) probing pulse could be timed to hit the molecular beam. For this second approach there are upcoming energy recovery Linac ('ERL') X-ray sources where the X-ray flux is not as high as the X-ray laser but where intrinsic few hundred fs pulse-length time-resolution per photon bunch should become available, e.g. as envisaged at CHESS, Cornell.

20

Overall summary and future thoughts

We have discussed the state-of-the-art in terms of crystallization and in particular the sample, emphasizing the practical aspects, screening and optimization. We have described how to overcome difficulties when high-quality crystals cannot be obtained, the current difficulties in tackling macromolecules associated with membranes and some lesser known alternative techniques. Although it is feasible to screen thousands of potential crystallization conditions, the crystallization problem has not been solved. The current state-of-the-art demonstrates that crystallization is still a bottleneck that requires further development; to be blunt, the wish to achieve crystals with certainty is still not a reality.

The outlook appears good, not least as there are promising developments in all stages of the process. New tools to monitor and control the crystallization process, such as dynamic light scattering (DLS) are now routinely employed. DLS allows the experimenters to pinpoint the exact timing of dilution or 'backing off' from nucleation to growth conditions by detecting the formation of aggregates or nuclei before they become visible under the microscope. Interferometry techniques have allowed monitoring of refraction-index changes in the mother liquor around growing crystals. The depletion zone around such a growing crystal, and its stabilization via various physical techniques to effect an optimal crystal perfection, has been established as a guiding concept.

The results of crystal-defect imaging techniques such as topography yield information on the crystal-growth process, which can be especially informative. Supplemented with mosaicity and reciprocal-space mapping, such 'combinational analyses' allow the detailed dissection of a history of a given crystal's growth. This affords lessons for improved crystal-growth avenues to be pursued. Ultraviolet light can be used to distinguish between salt and protein crystals thus indicating which 'hits' are worth pursuing further to the optimization stage. A fluorescence-based thermal shift assay is now applied to the case of high-throughput screening of protein stability as a function of the usual parameters, with obvious applicability to selection of final conditions.

In terms of analysis of the resulting crystal, we have then covered diffraction aspects from the basic studies to the effects of short and long-range ordering to cases where the resulting crystals are twinned or have other disorders. We have then covered how to get the best data from non-ideal systems and even when we have very tiny crystals to be studied singly or in the bulk as a crystalline powder. Progress in crystallographic capabilities at the 3rd-generation SR X-ray sources has enabled structural information to be extracted from cases that would have previously failed.

The diffraction properties of crystals stem from their order. Empirically, long-range order helps to improve signal-to-noise. We can measure the true long-range order using specialized techniques to minimize the geometrical and spectral properties already present in the X-ray beam. We can optimize 'regular' data collection in practice by matching this order ('mosaicity') with the X-ray beam smearing parameters and thus enhance the signal-to-noise notably at the 'resolution edge' of the measurable pattern. Similar principles apply to neutron-data measurement.

Twinning is a problem that occurs in crystals. As we have discussed in detail, considerable understanding of twinning in macromolecular crystals now exists including crystal-growth remediation possibilities or how to proceed with crystal-structure analysis in many cases if twinning can be remediated. Less is currently known about other lattice disorders, which we have also discussed, but understanding is improving and the number of examples increasing.

There are several techniques available to improve data collection. A common one is cryo-cooling, which substantially extends the crystal lifetime in the X-ray beam. This has made the MAD technique successful to the point of being 'routine' and allowed the full exploitation of third-generation synchrotron radiation sources. Another technique is controlled dehydration and humidity control. While not mitigating radiation damage this has had significant success in extending diffraction resolution in notable cases. The combination of the two techniques seems a practical direction to take for more recalcitrant crystal samples for the future.

Diffraction techniques and technologies are widening in scope and we are still seeing extensive and exciting developments. Ever more analyses of macromolecular, single microcrystals are a notable trend as X-ray microbeams ($\sim$10 micrometres typically at present) have become more widely available. Powder studies demonstrate that structural information can be obtained from the smallest crystallites gathered in bulk.

As we stated in our preface, 'the methods of today have a sense of maturity', but we believe there is much more to come. New and upgraded sources providing neutrons, synchrotron X-radiation and coherent fs X-ray pulses from free-electron lasers also promise further radical shifts in researcher expectations and horizons as well as methods in structural chemistry and biology.

Glossary of abbreviations, terms and symbols

Abbreviations

General

ADP	atomic displacement parameter
CCD	charge coupled device (used in X-ray and visible light imaging)
CMC	critical micellar concentration
DLS	dynamic light scattering
DPI	diffraction-component precision index
DQE	detector quantum efficiency
DVR/T	drop volume ratio/temperature (a form of optimization)
EF	electric field
FID	free interface diffusion (a form of crystallization)
HF	homogeneous field (magnetic)
ID	insertion device (sometimes used as a prefix to identify the type of synchrotron beamline)
IHF	inhomogeneous field (magnetic)
IUCr	International Union of Crystallography
LIMS	Laboratory Information Management System
LOT	large-angle oscillation technique (see Chapter 15)
MAD	multiple anomalous dispersion (a method to obtain phase information)
MIR	multiple isomorphous replacement (a method to obtain phase information)
MIRAS	multiple isomorphous replacement with anomalous scattering (see above)
NMR	nuclear magnetic resonance (a solution-based structural technique)
PEG	polyethylene glycol
QUELS	quasi-elastic light scattering (see DLS)
SAD	single-wavelength anomalous dispersion (a method to obtain phase information)
SAXS	small-angle X-ray scattering (a low-resolution solution-based technique)
SLS	static light scattering
S/N	single-to-noise ratio in a diffraction experiment
VD	vapour diffusion (a form of crystallization)
SR	synchrotron radiation
WAXS	wide-angle X-ray scattering (similar to SAXS but to a higher resolution)
XAS	X-ray absorption spectroscopy (a probe for metal binding environment)

XERLP	X-ray Energy Recovery Linac Project
XFEL	X-ray free-electron laser

Synchrotron sources

APS	Advanced Photon Source, Chicago, USA
CHESS	Cornell High Energy Synchrotron Source, Cornell, USA
DLS	Diamond Light Source, Rutherford Appleton Laboratory, UK
ESRF	European Synchrotron Radiation Facility, Grenoble, France
NSLS	National Synchrotron Light Source, Brookhaven, USA
SRS	Synchrotron Radiation Source, Daresbury Laboratory, UK
SSRL	Stanford Synchrotron Radiation Laboratory, USA

Neutron sources

ILL	Institut Laue Langevin, Grenoble, France
ISIS	Spallation Source, Rutherford Appleton Laboratory, UK
J-PARC	Japan Proton Accelerator Research Complex
SNS	Spallation Neutron Source, Oak Ridge, USA

TERMS

Crystallization

Cloud point: The point at which the dissolved sample is no longer completely soluble and a faint cloudiness becomes apparent in the previously clear solution.

Crystallization slot: Solution conditions where the individual macromolecules making up the crystal become slightly but not too attractive to one another.

Crystal growth (crystallization): Packing of macromolecules into an ordered symmetry related array starting from an initial nucleus.

High throughput: The application of usually automated techniques to perform many experiments sampling a diverse area of chemical, biological and physical space. Not necessarily high output.

Nucleation: The step when macromolecules separate in solution form clusters that may go on to growth. Nucleation can be either homogeneous, without the influence of foreign particles, or heterogeneous, with the influence of foreign particles.

Optimization: Fine tuning of crystallization conditions to take initially promising results to crystals suitable for diffraction analysis or to improve poorly diffracting crystals.

Phase diagram: A phase diagram for crystallization shows which state (liquid, crystalline or amorphous solid (precipitate)) is stable under a variety of crystallization parameters such as pH, temperature, etc. It describes the crystallization experiment and allows prediction of regions of conditions suitable to optimize further crystals.

Precipitation: The solid phase of a sample. This may be ordered precipitate, i.e. microcrystalline, or disordered from random aggregation.

Screening: Searching for conditions that indicate evidence of possible crystallization.

Solubility: Amount of a substance that will dissolve in a liquid at a given pH and temperature. The solubility is defined as the concentration of protein in the solute that is in equilibrium with crystals.

Supersolubility curve: The line on the protein crystallization phase diagram separating conditions where spontaneous nucleation (or phase separation, precipitation) occurs from conditions where the crystallization solution remains clear if left undisturbed.

Water activity: Defined as the ratio of the vapor pressure of water in a material to the vapor pressure of pure water at the same temperature and used as a measure describing the driving force for movement of water.

X-ray analysis

Atomic displacement factor (or B_{factor}): A measure of the effective diameter of an atom's electron density. Various kinds of disorder (static and thermal) effectively 'spread out' the electron density of a given atom, increasing the ADP. The B-factor is related to the rms error in an atom's position and is related to diffraction resolution.

Beamline: usually referred to in the synchrotron-source context, but also applies to neutron sources, with respect to the optics bringing the X-ray (or neutron beam) to the apparatus for measuring diffraction data from a sample.

Bragg's law: Introduced by Sir Lawrence Bragg, this describes the condition for a plane wave to be diffracted by a family of lattice planes $2d\sin\theta = n\lambda$ where d is the lattice spacing, θ the angle between the wavevector of the incident plane wave, and the lattice planes, λ its wavelength and n is an integer, known as the order of the reflection.

Crystal-structure analysis: First, using X-rays, the 3-dimensional layout of the arrangement of atoms and their individual chemical assignment is made. Where hydrogens are of particular interest the determination of the presence and positions of these is ideally done by a combination of 'ultrahigh-resolution X-ray analysis' and neutron macromolecular crystal-structure analysis.

Cryo-cooling: a method for the preservation of a sample, usually a single crystal but also a powder, in a cryo-preserved vitreous ice state. Found to be especially advantageous in X-ray diffraction to minimize X-radiation damage. It is also used in neutron studies (non-damaging as a probe), because it can reduce the atomic vibrations and enhance the diffraction resolution. With either X-ray or neutron or electron beams cryo-cooling can be used to 'freeze-trap' time-resolved structural intermediates down to time resolutions of $\sim$1 ms.

Miller indices, (h,k,l): used to describe a family of parallel planes that 'label' a diffraction reflection. It is a purely geometric concept and the planes need not include any atoms, but usually do.

Mosaicity: The width of the distribution of misorientation angles of all the unit cells in a crystal, usually described in terms of macroscopic 'domains' of the crystal. The measured mosaicity is a convolution of crystal and experimental properties. Typically, unless special conditions are set up, the crystal mosaicity is masked by geometric and spectral parameters of the beam.

Phase problem: Diffracted X-rays and neutrons obey Bragg's law. The diffracted radiation suffers a phase shift arising from the spatial distribution of individual atoms. The amplitudes are directly derivable from the experimental measured intensities of the diffracted beams, but the phases are not. Without a knowledge of the phases, it is not possible to reconstruct the individual atomic positions. This is the phase problem.

R factor: The reliability factor, which is a measure of the agreement between the amplitudes of the structure-factor amplitudes calculated from a crystallographic model and those from the original X-ray diffraction data. The R factor is calculated after each cycle of least-squares structure refinement to assess progress. The final R factor is one measure of final model structure quality. A related, very important, parameter is the R_{free} reliability factor calculated from a subset of reflections (e.g. 5%) which are retained outside of the main set of reflections, and give an independent measure of progress towards convergence to a final model.

Reflection: (Sometimes termed spot). See Miller indices.

Resolution limit: The resolution of a particular crystal's diffraction is typically given in Angstroms. It is the limit of the measured diffraction, i.e. shortest reciprocal-space distance sampled of all the reflections and in practice is often defined as where the average intensity / error on intensity ($<I/\sigma(I)>$) crosses 2.0.

Twinning: Twinning is the oriented association of two or more crystals of the same phase, which are related by an operation in direct space mapping the orientation of a crystal onto that of the other crystal(s) of the twin. Twins can form during crystal growth, following a growth transition, or as the result of a mechanical action.

SYMBOLS

Crystallization and crystal-growth-monitoring symbols

B	is the second virial coefficient
c	is the concentration of the protein

c/s	supersaturation of a solution
dn/dc	is the refractive index increment for the protein/solvent pair
Mw	is the weight-average molecular weight of the protein
n	is the refractive index of the solvent
N_A	is Avogadro's number (per mole)
R_{90}	is the excess Rayleigh factor at a scattering angle of 90 degrees
s	is the solubility of the protein

Diffraction symbols

A	is the fractional absorption by the crystal of the X-ray beam
α	is the twinning fraction of a crystal
$\delta\lambda/\lambda$	wavelength bandwidth
Δ_{radial}	is the reflection elongation in Laue diffraction in the radial direction
d_{min}	is the resolution limit of the diffraction data
E_{obs}, or I_{hkl}	is the observed intensity of the reflection
$\mathbf{F}$	is the structure factor
f_0	is the basic scattering factor of an atom
$f'(\lambda)$ and $f''(\lambda)$	are the real and imaginary components of the anomalous scattering (known as resonant scattering at an absorption edge)
γ_V and γ_H	are the vertical and horizontal angular divergence of the X-ray beam at the sample
I_o	is the incident intensity
h,k,l	Miller indices of reflection
λ	is the wavelength
L	is the Lorentz factor
ζ	is the position of the corresponding reciprocal-lattice point projected onto the rotation axis
η	is the intrinsic mosaicity of a crystal
P	is the polarization
R_{merge}	*is the* statistical agreement between the intensities of symmetry related reflections
$(S/N)_{\text{out}}/(S/N)_{\text{in}}$	is the detector quantum efficiency (DQE)
$\sigma(x, B_{\text{avg}})$	is the Cruickshank diffraction precision index being the typical standard deviation of a coordinate x for all atoms with average atomic displacement factor, or B Factor $B_{\text{avg.}}$
θ	is the Bragg angle of reflection
ϕ_R	is the measured reflection width
V_o	is the unit-cell volume
V_{cr}	is the illuminated volume of the crystal
XTD	is the crystal to detector distance
ω	is the angular rotation velocity of the crystal during measurements

References

(1987): ESRF Foundation Phase Report, ESRF, Grenoble, France.

(1994): The CCP4 suite: programs for protein crystallography. *Acta Crystallogr D Biol Crystallogr* **50**, 760–3.

Adachi, H., Umena, Y., Enami, I., Henmi, T., Kamiya, N., and Shen, J. R. (2009): Towards structural elucidation of eukaryotic photosystem II: Purification, crystallization and preliminary X-ray diffraction analysis of photosystem II from a red alga. *Biochim. Biophys. Acta Bioenerg.* **1787**, 121–128.

Adams, P. D., Gopal, K., Grosse-Kunstleve, R. W., Hung, L. W., Ioerger, T. R., McCoy, A. J., Moriarty, N. W., Pai, R. K., Read, R. J., Romo, T. D., Sacchettini, J. C., Sauter, N. K., Storoni, L. C., and Terwilliger, T. C. (2004): Recent developments in the PHENIX software for automated crystallographic structure determination. *J Synch Radiat* **11**, 53–5.

Adams, P. D., Grosse-Kunstleve, R. W., Hung, L. W., Ioerger, T. R., McCoy, A. J., Moriarty, N. W., Read, R. J., Sacchettini, J. C., Sauter, N. K., and Terwilliger, T. C. (2002): PHENIX: building new software for automated crystallographic structure determination. *Acta Crystallogr D Biol Crystallogr* **58**, 1948–54.

Ahmed, H. U., Blakeley, M. P., Cianci, M., Cruickshank, D. W. J., Hubbard, J. A., and Helliwell, J. R. (2007): The determination of protonation states in proteins. *Acta Crystallogr D Biol Crystallogr* **63**, 906–922.

Al-Haq, M. I., Lebrasseur, E., Choi, W. K., Tsuchiya, H., Torii, T., Yamazaki, H., and Shinohara, E. (2007a): An apparatus for electric-field-induced protein crystallization. *J Appl Crystallogr* **40**, 199–201.

Al-Haq, M. I., Lebrasseur, E., Tsuchiya, A., and Torii, T. (2007b): Protein crystallization under an electric field. *Crystallogr Rev* **13**, 29–64.

Altarelli, M. (1991): The European Synch Radiat Facility in Grenoble – a Progress Report. *Physica Scripta* **T34**, 73–76.

Amos, L. A., Jubb, J. S., Henderson, R., and Vigers, G. (1984): Arrangement of protofilaments in two forms of tubulin crystal induced by vinblastine. *J Mol Biol* **178**, 711–29.

Andrews, S. J., Hails, J. E., Harding, M. M., and Cruickshank, D. W. J. (1987): The mosaic spread of very small crystals deduced from Laue diffraction patterns. *Acta Crystallogr A* **43**, 70–73.

Andrews, S. J., Papiz, M. Z., Mcmeeking, R., Blake, A. J., Lowe, B. M., Franklin, K. R., Helliwell, J. R., and Harding, M. M. (1988): Piperazine Silicate (Eu-19) – the Structure of a Very Small Crystal Determined with Synch Radiat. *Acta Crystallogr Section B-Structural Science* **44**, 73–77.

Arai, S., Chatake, T., Suzuki, N., Mizuno, H., and Niimura, N. (2004): More rapid evaluation of biomacromolecular crystals for diffraction experiments. *Acta Crystallogr D Biol Crystallogr* **60**, 1032–9.

Arakali, S. V., Luft, J. R., and Detitta, G. T. (1995): Nonideality of Aqueous-Solutions of Polyethylene-Glycol – Consequences for Its Use as a Macromolecular Crystallizing Agent in Vapor-Diffusion Experiments. *Acta Crystallogr D Biol Crystallogr* **51**, 772–779.

Arndt, U. W. (1984): Optimum X-ray wavelength for protein crystallography. *J Appl Crystallogr* **17**, 118–119.

Asherie, N. (2004): Protein crystallization and phase diagrams. *Methods* **34**, 266–272.

Astier, J. P., Veesler, S., and Boistelle, R. (1998): Protein crystals orientation in a magnetic field. *Acta Crystallogr D Biol Crystallogr* **54**, 703–6.

Ataka, M. (1993): Protein Crystal-Growth – an Approach Based on Phase-Diagram Determination. *Phase Transitions* **45**, 205–219.

Ataka, M., Katoh, E., and Wakayama, N. I. (1997): Magnetic orientation as a tool to study the initial stage of crystallization of lysozyme. *J Cryst Growth* **173**, 592–596.

Azuma, T., Tsukamoto, K., and Sunagawa, I. (1989): Clustering Phenomenon and Growth Units in Lysozyme Aqueous-Solution as Revealed by Laser-Light Scattering Method. *J Cryst Growth* **98**, 371–376.

Backmark, A., Nyblom, M., Tornroth-Horsefield, S., Kosinska-Eriksson, U., Norden, K., Fellert, M., Kjellbom, P., Johanson, U., Hedfalk, K., Lindkvist-Petersson, K., Neutze, R., and Horsefield, R. (2008): Affinity tags can reduce merohedral twinning of membrane protein crystals. *Acta Crystallogr D Biol Crystallogr* **64**, 1183–1186.

Baldock, P., Mills, V., and Stewart, P. S. (1996): A comparison of microbatch and vapour diffusion for initial screening of crystallization conditions. *J Cryst Growth* **168**, 170–174.

Banci, L., Bertini, I., Cramaro, F., Del Conte, R., Rosato, A., and Viezzoli, M. S. (2000): Backbone dynamics of human Cu,Zn superoxide dismutase and of its monomeric F50E/G51E/E133Q mutant: the influence of dimerization on mobility and function. *Biochemistry* **39**, 9108–18.

Barber, J. (2009): Photosynthetic energy conversion: natural and artificial. *Chemical Society Reviews* **38**, 185–196.

Bard, J., Ercolani, K., Svenson, K., Olland, A., and Somers, W. (2004): Automated systems for protein crystallization. *Methods* **34**, 329–347.

Barker, A., I,, Southworth-Davies, R. J., Paithankar, K. S., Carmichael, I., and Garman, E. (2009): Room-temperature scavengers for macromolecular crystallography: increased lifetimes and modified dose dependence of the intensity decay. *J Synch Radiat* **16**, 205–216.

Bartels, W. J. (1983): Characterization of thin layers on perfect crystals with a mutlipurpose high resolution diffractometer. *J Vac Sci Technol* **B1**, 338–345.

Begg, G., and Speicher, D. (1999): Mass spectrometry detection and reduction of disulfide adducts between reducing agents and recombinant proteins with highly reactive cysteines. *J Biomol Tech* **10**, 17–20.

Bellamy, H. D., Snell, E. H., Lovelace, J., Pokross, M., and Borgstahl, G. E. (2000): The high-mosaicity illusion: revealing the true physical characteristics of macromolecular crystals. *Acta Crystallogr D Biol Crystallogr* **56**, 986–95.

Berejnov, V., Husseini, N. S., Alsaied, O. A., and Thorne, R. E. (2006): Effects of cryoprotectant concentration and cooling rate on vitrification of aqueous solutions. *J Appl Crystallogr* **39**, 244–251.

Berger, J. M. (2007): Soft polymer microfluidic applications to protein crystallization. In N. E. Chayen (ed.): *Protein crystallization strategies for structural genomics*, International University Line, La Jolla, California.

Bergfors, T. (2003): Seeds to crystals. *J Struct Biol* **142**, 66–76.

Bergfors, T. M. (1999): *Crystallization of proteins: techniques, strategies and tips.* International University Line.

Berisio, R., Lamzin, V. S., Sica, F., Zagari, A., Wilson, K. S., and Mazzarella, L. (1999). Protein titration in the crystal state. *J Mol Biol* **292**, 845–854.

Bern, M., Goldberg, D., Stevens, R. C., and Kuhn, P. (2004): Automatic classification of protein crystallization images using a curve-tracking algorithm. *J Appl Crystallogr* **37**, 279–287.

Bernal, J. D., and Crowfoot, D. (1934): X-ray photographs of crystalline pepsin. *Nature* **133**, 794–795.

Bernard, Y., Degoy, S., Lefaucheux, F., and Robert, M. C. (1994): A Gel-Mediated Feeding Technique for Protein Crystal-Growth from Hanging Drops. *Acta Crystallogr D Biol Crystallogr* **50**, 504–507.

Berry, I. M., Dym, O., Esnouf, R. M., Harlos, K., Meged, R., Perrakis, A., Sussman, J. L., Walter, T. S., Wilson, J., and Messerschmidt, A. (2006): SPINE high-throughput crystallization, crystal imaging and recognition techniques: current state, performance analysis, new technologies and future aspects. *Acta Crystallogr D Biol Crystallogr* **62**, 1137–1149.

Berthold, C. L., Sidhu, H., Ricagno, S., Richards, N. G., and Lindqvist, Y. (2006): Detection and characterization of merohedral twinning in crystals of oxalyl-coenzyme A decarboxylase from Oxalobacter formigenes. *Biochimica Et Biophysica Acta-Proteins and Proteomics* **1764**, 122–128.

Berthold, C. L., Toyota, C. G., Moussatche, P., Wood, M. D., Leeper, F., Richards, N. G., and Lindqvist, Y. (2007): Crystallographic snapshots of oxalyl-CoA decarboxylase give insights into catalysis by nonoxidative ThDP-dependent decarboxylases. *Structure* **15**, 853–61.

Beteva, A., Cipriani, F., Cusack, S., Delageniere, S., Gabadinho, J., Gordon, E. J., Guijarro, M., Hall, D. R., Larsen, S., Launer, L., Lavault, C. B., Leonard, G. A., Mairs, T., McCarthy, A., McCarthy, J., Meyer, J., Mitchell, E., Monaco, S., Nurizzo, D., Pernot, P., Pieritz, R., Ravelli, R. G. B., Rey, V., Shepard, W., Spruce, D., Stuart, D. I., Svensson, O., Theveneau, P., Thibault, X., Turkenburg, J., Walsh, M., and

McSweeney, S. M. (2006): High-throughput sample handling and data collection at synchrotrons: embedding the ESRF into the high-throughput gene-to-structure pipeline. *Acta Crystallogr D Biol Crystallogr* **62**, 1162–1169.

Biertumpfel, C., Basquin, J., Suck, D., and Sauter, C. (2002): Crystallization of biological macromolecules using agarose gel. *Acta Crystallogr D Biol Crystallogr* **58**, 1657–1659.

Blake, C. C. F., and Phillips, D. C. (1962): Effects of X-irradiaton on single crystals of myoglobin. Symposium on the Biological Effects of Ionising Radiation at the Molecular Level, pp. 183–191.

Blakeley, M. P., Cianci, M., Helliwell, J. R., and Rizkallah, P. J. (2004): Synchrotron and neutron techniques in biological crystallography. *Chem. Soc. Rev.* **33**, 548–557.

Blakeley, M. P., Langan, P., Niimura, N., and Podjarny, A. (2008): Neutron crystallography: opportunities, challenges, and limitations. *Curr Opin Struct Biol* **18**, 593–600.

Blow, D. M. (2002a): *Outline of crystallography for biologists*. Oxford University Press, Oxford, UK.

Blow, D. M. (2002b): Rearrangement of Cruickshank's formulae for the diffraction-component precision index. *Acta Crystallogr D Biol Crystallogr* **58**, 792–797.

Blundell, T. L., and Johnson, L. N. (1976): *Protein Crystallography*. Academic Press. New York.

Bogan, M. J. *et al.* (2008): Single particle X-ray diffractive imaging. *Nano Lett* **8**, 310–316.

Boggon, T. J., Chayen, N. E., Snell, E. H., Dong, J., Lautenschlager, P., Potthast, L., Siddons, D. P., Stojanoff, V., Gordon, E., Thompson, A. W., Zagalsky, P. F., Bi, R. C., and Helliwell, J. R. (1998): Protein crystal movements and fluid flows during microgravity growth. *Philos Trans. R. Soc. Lond. A* **356**, 1045–1061.

Boggon, T. J., Helliwell, J. R., Judge, R. A., Olczak, A., Siddons, D. P., Snell, E. H., and Stojanoff, V. (2000): Synchrotron X-ray reciprocal-space mapping, topography and diffraction resolution studies of macromolecular crystal quality. *Acta Crystallogr D Biol Crystallogr* **56**, 868–80.

Borgstahl, G. E. (2007): How to use dynamic light scattering to improve the likelihood of growing macromolecular crystals. *Methods Mol Biol* **363**, 109–29.

Borgstahl, G. E., Pokross, M., Chehab, R., Sekher, A., and Snell, E. H. (2000): Cryo-trapping the six-coordinate, distorted-octahedral active site of manganese superoxide dismutase. *J Mol Biol* **296**, 951–9.

Borgstahl, G. E. O., Vahedi-Fardi, A., Lovelace, J., Bellamy, H., and Snell, E. H. (2001): A test of macromolecular crystallization in microgravity: Large, well-ordered insulin crystals. *Acta Crystallogr D Biol Crystallogr* **57**, 1204–1207.

Bourenkov, G. P., and Popov, A. N. (2006): A quantitative approach to data-collection strategies. *Acta Crystallogr D Biol Crystallogr* **62**, 58–64.

Bourgeois, D., and Weik, M. (2009): Kinetic protein crystallography: a tool to watch proteins in action. *Crystallogr Rev* **15**, 87–118.

Bowen, D. K., and Tanner, B. K. (1998): *High resolution X-ray diffractometry and topography*. Taylor & Francis. Bristol PA.

Bowler, M. W., Montgomery, M. G., Leslie, A. G. W., and Walker, J. E. (2006): Reproducible improvements in order and diffraction limit of crystals of bovine mitochondrial F-1-ATPase by controlled dehydration. *Acta Crystallogr D Biol Crystallogr* **62**, 991–995.

Bragg, W. L., and Howells, E. R. (1954): X-ray Diffraction by Imidazole Methaemoglobin. *Acta Crystallogr* **7**, 409–411.

Breyer, W. A., Kingston, R. L., Anderson, B. F., and Baker, E. N. (1999): On the molecular-replacement problem in the presence of merohedral twinning: structure of the N-terminal half-molecule of human lactoferrin. *Acta Crystallogr D Biol Crystallogr* **55**, 129–138.

Britton, D. (1972): Estimation of Twinning Parameter for Twins with Exactly Superimposed Reciprocal Lattices. *Acta Crystallogr Section A-Crystal Physics Diffraction Theoretical and General Crystallography* **28**, 296–297.

Brockhouse, B. N. (1955): Energy distribution of neutrons scattered by paramagnetic substances. *Phys. Rev.* **99**, 601–603.

Brzozowski, A. M., and Walton, J. (2001): Clear strategy screens for macromolecular crystallization. *J Appl Crystallogr* **34**, 97–101.

Budayova-Spano, M., Dauvergne, F., Audiffren, M., Bactivelane, T., and Cusack, S. (2007): A methodology and an instrument for the temperature-controlled optimization of crystal growth. *Acta Crystallogr D Biol Crystallogr* **63**, 339–347.

Burmeister, W. P. (2000): Structural changes in a cryo-cooled protein crystal owing to radiation damage. *Acta Crystallogr D Biol Crystallogr* **56**, 328–341.

Burning, J. H., Herriot, D. R., Gallagher, J. E., Rosenfeld, D. P., and White, A. D. (1974): Digital wavefront measuring interferometer for testing optical surfaces and lenses. *Appl. Opt.* **13**, 2693.

Buts, L., Dao-Thi, M. H., Wyns, L., and Loris, R. (2004): Untangle, a tool for filtering overlapping diffraction patterns from multicrystals. *Acta Crystallogr D Biol Crystallogr* **60**, 983–984.

Caffrey, M. (2003): Membrane protein crystallization. *J Struct Biol* **142**, 108–132.

Canaves, J. M., Page, R., Wilson, I. A., and Stevens, R. C. (2004): Protein biophysical properties that correlate with crystallization success in Thermotoga maritima: maximum clustering strategy for structural genomics. *J Molec Biol* **344**, 977–91.

Carpenter, E. P., Beis, K., Cameron, A. D., and Iwata, S. (2008): Overcoming the challenges of membrane protein crystallography. *Curr Opin Struct Biol* **18**, 581–6.

Carter, C. W. (1997): Response surface methods for optimizing and improving reproducibility of crystal growth. *Macromolec Crystallogr, Pt A* **276**, 74–99.

Carter, C. W., and Carter, C. W. (1979): Protein Crystallization Using Incomplete Factorial-Experiments. *J Biol Chem* **254**, 2219–2223.

Carter, D. C., Rhodes, P., McRee, D. E., Tari, L. W., Dougan, D. R., Snell, G., Abola, E., and Stevens, R. C. (2005): Reduction in diffuso-convective disturbances in nanovolume protein crystallization experiments. *J Appl Crystallogr* **38**, 87–90.

Carugo, O., and Carugo, K. D. (2005): When X-rays modify the protein structure: radiation damage at work. *Trends Biochem Sci* **30**, 213–219.

Cassetta, A., Deacon, A. M., Ealick, S. E., Helliwell, J. R., and Thompson, A. W. (1999): Development of instrumentation and methods for MAD and structural genomics at the SRS, ESRF, CHESS and Elettra facilities. *J Synch Radiat* **6**, 822–833.

Catti, M., and Ferraris, G. (1976): Twinning by Merohedry and X-Ray Crystal-Structure Determination. *Acta Crystallogr Section A* **32**, 163–165.

Cavanagh, J., Fairbrother, W. J., Palmer III, A. G., Skelton, N. J., and Rance, M. (2006): *Protein NMR spectroscopy: principles and practice* Academic Press.

Cernik, R. J., Cheetham, A. K., Prout, C. K., Watkin, D. J., Wilkinson, A. P., and Willis, B. T. M. (1991): The Structure of Cimetidine ($C_{10}H_{16}N_6S$) Solved from Synchrotron-Radiation X-Ray-Powder Diffraction Data. *J Appl Crystallogr* **24**, 222–226.

Chandra, N., Acharya, K. R., and Moody, P. C. E. (1999): Analysis and characterization of data from twinned crystals. *Acta Crystallogr D Biol Crystallogr* **55**, 1750–1758.

Chao, W. C., Collins, J., Wang, S. W., Li, G. P., Bachman, M., and Lee, A. P. (2004): Control of concentration and volume gradients in microfluidic droplet arrays for protein crystallization screening. *Conf Proc IEEE Eng Med Biol Soc* **4**, 2623–6.

Chapman, H., Barty, A., Beetz, T., Cui, C., He, H., Howells, M., Marchesini, S., Noy, A., Rosen, R., Spence, J., Weierstall, U., Beetz, T., Jacobsen, C., and Shapiro, D. (2006): High resolution three-dimensional X-ray diffraction microscopy. *J Opt Soc Am* **23**, 1179–1200.

Chapman, H. N., Barty, A., Bogan, M. J., Boutet, S., Frank, M., Hau-Riege, S. P., Marchesini, S., Woods, B. W., Bajt, S., Benner, W. H., London, R. A., Plönjes, E., Kuhlmann, M., Treusch, R., Düsterer, S., Tschentscher, T., Schneider, J. R., Spiller, E., Möller, T., Bostedt, C., Hoener, M., Shapiro, D. A., Hodgson, K. O., van der Spoel, D., Burmeister, F., Bergh, M., Caleman, C., Huldt, G., Seibert, M. M., Maia, F., Lee, R. W., Szöke, A., Timneanu, N., and Hajdu, J. (2006): Femtosecond diffractive imaging with a soft-X-ray free-electron laser. *Nature Phys* **5**, 839–843.

Chayen, N. E. (1997): The role of oil in macromolecular crystallization. *Structure* **5**, 1269–1274.

Chayen, N. E. (1998): Comparative studies of protein crystallization by vapour-diffusion and microbatch techniques. *Acta Crystallogr D Biol Crystallogr* **54 (Pt 1)**, 8–15.

Chayen, N. E. (2003): Protein crystallization for genomics: throughput versus output. *J Struct Funct Genomics* **4**, 115–20.

Chayen, N. E. (2004): Turning protein crystallization from an art into a science. *Cur. Opin Struc. Biol* **14**, 577–583.

Chayen, N. E. (2005): Protein Crystallization: Automation, Robotization and Miniaturization, Chapter 3, pp. 29–48. In M. Sundstrum, M. Norin, and A. Edwards (eds.): *Structural genomics and high throughput structural biology*, CRC Press Taylor and Francis, Abingdon, UK.

Chayen, N. E. (2009): Oils for screening and optimization, Chapter 4, pp. 65–81. In T. Bergfors (ed.): *Protein crystallization*, International University Line, La Jolla, CA, USA.

Chayen, N. E., Radcliffe, J. W., and Blow, D. M. (1993): Control of Nucleation in the Crystallization of Lysozyme. *Protein Sci* **2**, 113–118.

Chayen, N. E., and Saridakis, E. (2002): Protein crystallization for genomics: towards high-throughput optimization techniques. *Acta Crystallogr D Biol Crystallogr* **58**, 921–927.

Chayen, N. E., and Saridakis, E. (2008): Protein crystallization: from purified protein to diffraction-quality crystal. *Nature Methods* **5**, 147–153.

Chayen, N. E., Saridakis, E., and Sear, R. P. (2006): Experiment and theory for heterogeneous nucleation of protein crystals in a porous medium. *Proc Nat Acad Sci USA* **103**, 597–601.

Chayen, N. E., Shaw Stewart, P. D., and Blow, D. M. (1992): Microbatch crystallization under oil — a new technique allowing many small-volume crystallization trials *J Cryst Growth* **122**, 176–180.

Chayen, N. E., Stewart, P. D. S., and Baldock, P. (1994): New Developments of the Impax Small-Volume Automated Crystallization System. *Acta Crystallogr D Biol Crystallogr* **50**, 456–458.

Chayen, N. E., Stewart, P. D. S., Maeder, D. L., and Blow, D. M. (1990): An Automated-System for Microbatch Protein Crystallization and Screening. *J Appl Crystallogr* **23**, 297–302.

Chen, L., Oughtred, R., Berman, H. M., and Westbrook, J. (2004): TargetDB: a target registration database for structural genomics projects. *Bioinformatics* **20**, 2860–2.

Chen, P. S., Shlichta, P. J., Wilcox, W. R., and Lefever, R. A. (1979): Convection phenomena during the growth of sodium chlorate crystals from solution. *J. Cryst Growth* **47**, 43–60.

Cherezov, V., and Caffrey, M. (2003): Nano-volume plates with excellent optical properties for fast, inexpensive crystallization screening of membrane proteins. *J Appl Crystallogr* **36**, 1372–1377.

Cherezov, V., Peddi, A., Muthusubramaniam, L., Zheng, Y. F., and Caffrey, M. (2004): A robotic system for crystallizing membrane and soluble proteins in lipidic mesophases. *Acta Crystallogr D Biol Crystallogr* **60**, 1795–1807.

Cherezov, V., Rosenbaum, D. M., Hanson, M. A., Rasmussen, S. G. F., Thian, F. S., Kobilka, T. S., Choi, H.-J., Kuhn, P., Weis, W. I., Kobilka, B. K., and Stevens, R. C. (2007): High-Resolution Crystal Structure of an Engineered Human 2-Adrenergic G Protein Coupled Receptor. *Science* **318**, 1258–1265.

Chernov, A. A. (1998): Crystal growth and crystallography. *Acta Crystallogr Section A* **54**, 859–872.

Chernov, A. A. (2003): Protein crystals and their growth. *J Struct Biol* **142**, 3–21.

Chernov, P. G., and Chernov, A. A. (2002): The physics of protein crystallization. *Solid State Phys* **57**, 1–147.

Chinte, U., Shah, B., DeWitt, K., Kirschbaum, K., Pinkerton, A. A., and Schall, C. (2005): Sample size: an important parameter in flash-cooling macromolecular crystallization solutions. *J Appl Crystallogr* **38**, 412–419.

Cianci, M., Antonyuk, S., Bliss, N., Bailey, M. W., Buffey, S. G., Cheung, K. C., Clarke, J. A., Derbyshire, G. E., Ellis, M. J., Enderby, M. J., Grant, A. F., Holbourn, M. P., Laundy, D., Nave, C., Ryder, R., Stephenson, P., Helliwell, J. R., and Hasnain, S. S. (2005a): A high-throughput structural biology/proteomics beamline at the SRS on a new multipole wiggler. *J Synch Radiat* **12**, 455–66.

Cianci, M., Helliwell, J. R., Helliwell, M., Kaucic, V., Logar, N. Z., Mali, G., and Tusar, N. N. (2005b): Anomalous scattering in structural chemistry and biology. *Crystallogr Rev* **11**, 245–335.

Cianci, M., Helliwell, J. R., and Suzuki, A. (2008): The interdependence of wavelength, redundancy and dose in sulfur SAD experiments. *Acta Crystallogr D Biol Crystallogr* **64**, 1196–1209.

Clancy, L. L., Finzel, B. C., Yem, A. W., Deibel, M. R., Jr., Strakalaitis, N. A., Brunner, D. P., Sweet, R. M., and Einspahr, H. M. (1994): Initial crystallographic analysis of a recombinant human interleukin-1 receptor antagonist protein. *Acta Crystallogr D Biol Crystallogr* **50**, 197–201.

Clay, M. D., Jenney, F. E., Hagedoorn, P. L., George, G. N., Adams, M. W. W., and Johnson, M. K. (2002): Spectroscopic studies of Pyrococcus furiosus superoxide reductase: Implications for active-site structures and the catalytic mechanism. *J Am Chem Soc* **124**, 788–805.

Cochran, W., and Howells, E. R. (1954): X-ray Diffraction by a Layer Structure Containing Random Displacements. *Acta Crystallogr* **7**, 412–415.

Cohen, A. E., Ellis, P. J., Miller, M. D., Deacon, A. M., and Phizackerley, R. P. (2002): An automated system to mount cryo-cooled protein crystals on a synchrotron beamline, using compact sample cassettes and a small-scale robot. *J Appl Crystallogr* **35**, 720–726.

Cohen, A. E., McPhillips, S. E., Song, J., and Miller, M. D. (2005): Automation of High-Throughput Protein Crystal Screening at SSRL. *Synch Radiat News* **18**, 28–35.

Colapietro, M., Cappuccio, G., Marciante, C., Pifferi, A., Spagna, R., and Helliwell, J. R. (1992): The X-ray diffraction station at the ADONE wiggler facility: preliminary results (including crystal perfection). *J Appl Crystallogr* **25**, 192–194.

Cork, C., O'Neill, J., Taylor, J., and Earnest, T. (2006): Advanced beamline automation for biological crystallography experiments. *Acta Crystallogr D Biol Crystallogr* **62**, 852–858.

Cosier, J., and Glazer, M. (1986): A Nitrogen-gas-stream cryostat for general X-ray diffraction studies. *J Appl Crystallogr* **19**, 105–107.

Cox, M. J., and Wever, P. C. (1987): Experiments with automated protein crystallization. *J Appl Crystallogr* **20**, 366–373.

Crennell, S., Takimoto, T., Portner, A., and Taylor, G. (2000): Crystal structure of the multifunctional paramyxovirus hemagglutinin-neuraminidase. *Nature Struct Biol* **7**, 1068–74.

Cruickshank, D. W. J. (1999): Remarks about protein structure precision. *Acta Crystallogr D Biol Crystallogr* **55**, 583–601.

Cruickshank, D. W. J., Carr, P. D., and Harding, M. M. (1992): Estimation of d_{min}, λ_{min} and λ_{max} from the Gnomonic Projections of Laue Patterns. *J Appl Crystallogr* **25**, 285–293.

Cudney, B., Patel, S., and McPherson, A. (1994a): Crystallization of Macromolecules in Silica-Gels. *Acta Crystallogr D Biol Crystallogr* **50**, 479–483.

Cudney, B., Patel, S., Weisgraber, K., and Newhouse, Y. (1994b): Screening and Optimization Strategies for Macromolecular Crystal-Growth. *Acta Crystallogr D Biol Crystallogr* **50**, 414–4123.

Cumbaa, C., and Jurisica, I. (2005): Automatic classification and pattern discovery in high-throughput protein crystallization trials. *J Struct Funct Genomics* **6**, 195–202.

Cumbaa, C. A., Lauricella, A., Fehrman, N., Veatch, C., Collins, R., Luft, J., DeTitta, G., and Jurisica, I. (2003): Automatic classification of sub-microlitre protein-crystallization trials in 1536-well plates. *Acta Crystallogr D Biol Crystallogr* **59**, 1619–1627.

Czepas, J., Devedjiev, Y., Krowarsch, D., Derewenda, U., Otlewski, J., and Derewenda, Z. S. (2004): The impact of Lys>Arg surface mutations on the crystallization of the globular domain of RhoGDI. *Acta Crystallogr D Biol Crystallogr* **60**, 275–80.

D'Arcy, A. (1994): Crystallizing Proteins – a Rational Approach. *Acta Crystallogr D Biol Crystallogr* **50**, 469–471.

D'Arcy, A., Frederic, V. A., and Marsh, M. (2007): An automated microseed matrix-screening method for protein crystallization. *Acta Crystallogr D Biol Crystallogr* **63**, 550–554.

D'Arcy, A., Mac Sweeney, A., and Habera, A. (2004): Modified microbatch and seeding in protein crystallization experiments. *J Synch Radiat* **11**, 24–6.

D'Arcy, A., Stihle, M., Kostrewa, D., and Dale, G. (1999): Crystal engineering: a case study using the 24 kDa fragment of the DNA gyrase B subunit from Escherichia coli. *Acta Crystallogr D Biol Crystallogr* **55**, 1623–5.

Darwin, C. G. (1922): The reflection of X-rays from imperfect crystals. *Philos Mag* **43**, 800–829.

Dauter, Z. (2003): Twinned crystals and anomalous phasing. *Acta Crystallogr D Biol Crystallogr* **59**, 2004–2016.

David, W. I. F., Shankland, K., and Shankland, N. (1998): Routine determination of molecular crystal structures from powder diffraction data. *Chem Commun*, 931–932.

Deisenhofer, J., Epp, O., Miki, K., Huber, R., and Michel, H. (1984): X-ray structure analysis of a membrane protein complex. Electron density map at 3 Å resolution and a model of the chromophores of the photosynthetic reaction center from Rhodopseudomonas viridis. *J Molec Biol* **180**, 385–98.

DeLucas, L., Long, M. M., Morre, K. M., Rosenblum, W. M., Bray, T. L., Smith, C., Carson, M., Narayana, S. V. L., Harrington, M. D., Carter, D., Clark, A. D. J., Nanni, R. G., Ding, J., Jacobo-Molina, A., Kamer, G., Hughes, S. H., Arnold, E., Einspahr, H. M., Clancy, L. L., Rao, G. S. J., Cook, P. F., Harris, B. G., Munson, S. H., Finzel, B. C., McPherson, A., Weber, P. C., Lewandowski, F. A., Nagabhushan, T. L., Trotta, P. P., Reichart, P., Navia, K. P., Wilson, K. P., Thompson, J. A., Richsrds, R. N., Bowersox, K. D., Meade, C. J., Baker, E. S., Bishop, S. P., B.J., D., Trinh, E., Prahl, J., Sacco, A. J., and Bugg, C. E. (1994): Recent trends and new hardware developments for protein crystal growth in microgravity. *J. Cryst Growth* **135**, 183–195.

Delucas, L. J., Hamrick, D., Cosenza, L., Nagy, L., McCombs, D., Bray, T., Chait, A., Stoops, B., Belgovskiy, A., William Wilson, W., Parham, M., and Chernov, N. (2005a): Protein crystallization: virtual screening and optimization. *Prog Biophys Mol Biol* **88**, 285–309.

DeLucas, L. J., Hamrick, D., Cosenza, L., Nagy, L., McCombs, D., Bray, T., Chait, A., Stoops, B., Belgovskiy, A., Wilson, W. W., Parham, M., and Chernov, N. (2005b): Protein crystallization: virtual screening and optimization. *Prog Biophys Mol Biol* **88**, 285–309.

DeLucas, L. J., Smith, C. D., Smith, H. W., Vijay-Kumar, S., Senadhi, S. E., Ealick, S. E., Carter, D. C., Snyder, R. S., Weber, P. C., and Salemme, F. R. (1989): Protein crystal growth in microgravity. *Science* **246**, 651–4.

DeLucas, L. J., Smith, C. D., Smith, W., Vijay-Kumar, S., Senadhi, S. E., Ealick, S. E., Carter, D. C., Snyder, R. S., Weber, P. C., Salemme, F. R., Ohlendorf, D. H., Einspahr, H. M., Clancy, L. L., Navia, M. A., McKeever, B. M., Nagabhushan, T. L., Nelson, G., McPherson, A., Koszelak, S., Taylor, G., Stammers, D., Powell, K., Darby, G., and Bugg, C. E. (1991): Protein crystal growth results for shuttle flights STS-26 and STS-29. *J. Cryst Growth* **110**, 302–311.

Derewenda, Z. S. (2004a): Rational protein crystallization by mutational surface engineering. *Structure (Camb)* **12**, 529–35.

Derewenda, Z. S. (2004b): The use of recombinant methods and molecular engineering in protein crystallization. *Methods* **34**, 354–63.

Derewenda, Z. S., and Vekilov, P. G. (2006): Entropy and surface engineering in protein crystallization. *Acta Crystallogr D Biol Crystallogr* **62**, 116–24.

Dierks, K., Meyer, A., Einspahr, H., and Betzel, C. (2008): Dynamic light scattering in protein crystallization droplets: Adaptations for analysis and optimization of crystallization processes. *Cryst Growth & Design* **8**, 1628–1634.

Dimmeler, E., Vossen, O., and Schroder, R. R. (2000): Determination of lattice-transform density profiles for multilayered three-dimensional microcrystals in electron crystallography. *J Appl Crystallogr* **33**, 1102–1112.

Dobbek, H., Gremer, L., Kiefersauer, R., Huber, R., and Meyer, O. (2002): Catalysis at a dinuclear [CuSMo(==O)OH] cluster in a CO dehydrogenase resolved at 1.1 Å resolution. *Proc Nat Acad Sci USA* **99**, 15971–15976.

Dobrianov, I., Finkelstein, K. D., Lemay, S. G., and Thorne, R. E. (1998): X-ray Topographic Studies of Protein Crystal Perfection and Growth. *Acta Crystallogr D Biol Crystallogr* **54**, 922–937.

Donadini, R., Wahlberga, M., Kohsaka, T., Ito, Y., and Fields, B. A. (2003): Crystallization and preliminary X-ray analysis of Yersinia pseudotuberculosis-derived mitogen. *Acta Crystallogr D Biol Crystallogr* **59**, 1330–1332.

Dong, A., Xu, X., Edwards, A. M., Chang, C., Chruszcz, M., Cuff, M., Cymborowski, M., Di Leo, R., Egorova, O., Evdokimova, E., Filippova, E., Gu, J., Guthrie, J., Ignatchenko, A., Joachimiak, A., Klostermann, N., Kim, Y., Korniyenko, Y., Minor, W., Que, Q., Savchenko, A., Skarina, T., Tan, K., Yakunin, A., Yee, A., Yim, V., Zhang, R., Zheng, H., Akutsu, M., Arrowsmith, C., Avvakumov, G. V., Bochkarev, A., Dahlgren, L. G., Dhe-Paganon, S., Dimov, S., Dombrovski, L., Finerty, P., Jr., Flodin, S., Flores, A., Graslund, S., Hammerstrom, M., Herman, M. D., Hong, B. S., Hui, R., Johansson, I., Liu, Y., Nilsson, M., Nedyalkova, L., Nordlund, P., Nyman, T., Min, J., Ouyang, H., Park, H. W., Qi, C., Rabeh, W., Shen, L., Shen, Y., Sukumard, D., Tempel, W., Tong, Y., Tresagues, L., Vedadi, M., Walker, J. R., Weigelt, J., Welin, M., Wu, H., Xiao, T., Zeng, H., and Zhu, H. (2007): In situ proteolysis for protein crystallization and structure determination. *Nature Methods* **4**, 1019–1021.

Dong, J., Boggon, T. J., Chayen, N. E., Raftery, J., Bi, R. C., and Helliwell, J. R. (1999): Bound-solvent structures for microgravity-, ground control-, gel- and microbatch-grown hen egg-white lysozyme crystals at 1.8 Å resolution. *Acta Crystallogr D Biol Crystallogr* **55**, 745–52.

Dornberger-Schiff, K. (1979): OD Structures, – a Game and a Bit more. *Kristall und Technik* **14**, 1027–1045.

Dorset, D. L., and Parsons, D. F. (1975): Electron-Diffraction from Single, Fully-Hydrated, Ox Liver Catalase Microcrystals. *Acta Crystallogr Section A* **A 31**, 210–215.

Drenth, J. (1999): *Principles of protein X-ray crystallography*. Springer-Verlag, New York, USA.

Dubnovitsky, A. P., Ravelli, R. B. G., Popov, A. N., and Papageorgiou, A. C. (2005): Strain relief at the active site of phosphoserine aminotransferase induced by radiation damage. *Protein Sci* **14**, 1498–1507.

Dubois, L., Jacquamet, L., Pecaut, J., and Latour, J. M. (2006): X-Ray photoreduction of a di(mu-oxo)(MnMnIV)-Mn-III complex occurs at temperatures as low as 20 K. *Chem Commun*, 4521–4523.

Ducruix, A., and Giege, R. (1999): *Crystallization of nucleic acids and proteins. A practical approach*. Oxford University Press, Oxford, UK.

Dunlop, K. V., and Hazes, B. (2005): A modified vapor-diffusion crystallization protocol that uses a common dehydrating agent. *Acta Crystallogr D Biol Crystallogr* **61**, 1041–1048.

Ďurovič, S. (2004): Layer stacking in general polytypic structures., pp. 752–765. In A. J. C. Wilson, and E. Prince (ed.): *International tables for crystallography.*, Kluwer Academic Publishers, Dordrecht / Boston / London.

Edwards, C., Palmer, S. B., Emsley, P., Helliwell, J. R., Glover, I. D., Harris, G. W., and Moss, D. S. (1990): Thermal Motion in Protein Crystals Estimated Using Laser-Generated Ultrasound and Young Modulus Measurements. *Acta Crystallogr Section A* **46**, 315–320.

Efremov, R., Moukhametzianov, R., Buldt, G., and Gordeliy, V. (2004): Physical detwinning of hemihedrally twinned hexagonal crystals of bacteriorhodopsin. *Biophys J* **87**, 3608–3613.

Einspahr, H., Suguna, K., Suddath, F. L., Ellis, G., Helliwell, J. R., and Papiz, M. Z. (1985): The Location of Manganese and Calcium-Ion Cofactors in Pea Lectin Crystals by Use of Anomalous Dispersion and Tunable Synchrotron X-Radiation. *Acta Crystallogr Section B-Struct Sci* **41**, 336–341.

Ellis, M. J., Antonyuk, S., and Hasnain, S. S. (2002): Resolution improvement from 'in situ annealing' of copper nitrite reductase crystals. *Acta Crystallogr D Biol Crystallogr* **58**, 456–8.

Engel, M., Hoffmann, T., Manhart, S., Heiser, U., Chambre, S., Huber, R., Demuth, H. U., and Bode, W. (2006): Rigidity and flexibility of dipeptidyl peptidase IV: Crystal structures of and docking experiments with DPIV. *J Molec Biol* **355**, 768–783.

Engel, M., Hoffmann, T., Wagner, L., Wermann, M., Heiser, U., Kiefersauer, R., Huber, R., Bode, W., Demuth, H. U., and Brandstetter, H. (2003): The crystal structure of dipeptidyl peptidase IV(CD26) reveals its functional regulation and enzymatic mechanism. *Proc Nat Acad Sci USA* **100**, 5063–5068.

Esnouf, R. M., Ren, J. S., Garman, E. F., Somers, D. O., Ross, C. K., Jones, E. Y., Stammers, D. K., and Stuart, D. I. (1998): Continuous and discontinuous changes in the unit cell of HIV-1 reverse transcriptase crystals on dehydration. *Acta Crystallogr D Biol Crystallogr* **54**, 938–953.

Estebanez-Perpina, E., Fuentes-Prior, P., Belorgey, D., Braun, M., Kiefersauer, R., Maskos, K., Huber, R., Rubin, H., and Bode, W. (2000): Crystal structure of the caspase activator human granzyme B, a proteinase highly specific for an Asp-P1 residue. *Biological Chemistry* **381**, 1203–1214.

Evdokimov, A. (2007): Rescuing low-quality crystals: *Molecular dimensions monograph series*, Molecular Dimensions Ltd., UK.

Ferre-D'Amare, A. R., and Burley, S. K. (1994): Use of dynamic light scattering to assess crystallizability of macromolecules and macromolecular assemblies. *Structure* **2**, 357–9.

Ferreira, K. N., Iverson, T. M., Maghlaoui, K., Barber, J., and Iwata, S. (2004): Architecture of the photosynthetic oxygen-evolving center. *Science* **303**, 1831–8.

Ferrer, J.-L., and Roth, M. (1998): An Algorithm for Automatic Mosaic Spread Analysis of a Protein Crystal. *J Appl Crystallogr* **33**, 433–436.

Fewster, P. F. (1996): Reciprocal space mapping, pp. 269–287. In A. Authier, S. Lagomarsino, and B. K. Tanner (ed.): *X-ray and neutron diffraction dynamical diffraction: Theory and applications*, Plenum Press, New York.

Fewster, P. F. (1997): Reciprocal space mapping. *Critical reviews in solid state and materials science* **22**, 69–110.

Fienup, J. R. (1982): *Phase retrieval algorithms: a comparison. Appl Opt* **21**, 2758–2769.

Fischetti, R. F., Xu, S. L., Yoder, D. W., Becker, M., Nagarajan, V., Sanishvili, R., Hilgart, M. C., Stepanov, S., Makarov, O., and Smith, J. L. (2009): Mini-beam collimator enables microcrystallography experiments on standard beamlines. *J Synch Radiat* **16**, 217–225.

Fisher, R. G., and Sweet, R. M. (1980): Treatment of Diffraction Data from Protein Crystals Twinned by Merohedry. *Acta Crystallogr Section A* **36**, 755–760.

Fisher, S. J., Helliwell, J. R., Khurshid, S., Govada, L., Redwood, C., Squire, J. M., and Chayen, N. E. (2008): An investigation into the protonation states of the C1 domain of cardiac myosin-binding protein C. *Acta Crystallogr D Biol Crystallogr* **64**, 658–64.

Forstner, M. (2000): SAXS, SANS and X-ray crystallography as complementary methods in the study of biological form and function. *J Appl Crystallogr* **33**, 519–523.

Fourme, R., Ducruix, A., Ries-Kautt, M., and Capelle, B. (1995): The Perfection of Protein Crystals Probed by Direct Recording of Bragg Reflection Profiles with a Quasi-Planar X-ray Wave. *J. Synch Radiat* **2**, 136–142.

Fourme, R., Ducruix, A., Ries-Kautt, M., and Capelle, B. (1999): Probing the quality of crystals of biological macromolecules using maximum resolution of diffraction data, Bragg reflection profiles and X-ray topographs. *J. Cryst Growth* **196**, 535–545.

Fowlis, W. W., Delucas, L. J., Twigg, P. J., Howard, S. B., Meehan, E. J., and Baird, J. K. (1988): Experimental and Theoretical-Analysis of the Rate of Solvent Equilibration in the Hanging Drop Method of Protein Crystal-Growth. *J Cryst Growth* **90**, 117–129.

Fox, K. M., and Karplus, P. A. (1993): Crystallization of Old Yellow Enzyme illustrates an effective strategy for increasing protein crystal size. *J Mol Biol* **234**, 502–7.

Friedel, G. (1926): *Leçons de cristallographie*. Berger-Levrault. Nancy, Paris, Strasbourg.

Friedel, G. (1933): Sur un nouveau type de macles. *Bull Soc Fr Miner Crist* **56**, 262–274.

Fung, R., Schneerson, V., Saldin, D., and Ourmazd, A. (2009): Structure from fleeting illumination of faint spinning objects in flight. *Nature Phys* **5**, 64–67.

Gamble, T. R., Clauser, K. R., and Kossiakoff, A. A. (1994): The Production and X-Ray Structure Determination of Perdeuterated Staphylococcal Nuclease. *Biophys Chem* **53**, 15–25.

Garavito, R. M., and Ferguson-Miller, S. (2001): Detergents as tools in membrane biochemistry. *J Biol Chem* **276**, 32403–6.

Garcia-Ruiz, J. M. (1994): Investigations on protein crystal growth by the gel acupuncture method. *Acta Crystallogr D Biol Crystallogr* **50**, 484–90.

Garcia-Ruiz, J. M. (2003a): Counterdiffusion methods for macromolecular crystallization. *Methods Enzymol* **368**, 130–54.

Garcia-Ruiz, J. M. (2003b): Nucleation of protein crystals. *J Struct Biol* **142**, 22–31.

Garcia-Ruiz, J. M., Gonzalez-Ramirez, L. A., Gavira, J. A., and Otalora, F. (2002): Granada Crystallisation Box: a new device for protein crystallisation by counter-diffusion techniques. *Acta Crystallogr D Biol Crystallogr* **58**, 1638–1642.

Garcia-Ruiz, J. M., Hernandez-Hernandez, A., Lopez-Jaramillo, J., and Thomas, B. (2001): Crystallization screening directly in electrophoresis gels. *J Cryst Growth* **232**, 596–602.

Garman, E. (1999): Cool data: quantity AND quality. *Acta Crystallogr D Biol Crystallogr* **55**, 1641–1653.

Garman, E., and Nave, C. (2002): Radiation damage to crystalline biological molecules: current view. *J Synch Radiat* **9**, 327–328.

Garman, E. F., and Doublie, S. (2003): Cryocooling of macromolecular crystals: optimization methods. *Methods Enzymol* **368**, 188–216.

Garman, E. F., and Mitchell, E. P. (1996): Glycerol concentrations required for cryoprotection of 50 typical protein crystallization solutions. *J Appl Crystallogr* **29**, 584–587.

Garman, E. F., and Nave, C. (2009): Radiation damage in protein crystals examined under various conditions by different methods. *J Synch Radiat* **16**, 129–32.

Garman, E. F., and Owen, R. L. (2006): Cryocooling and radiation damage in macromolecular crystallography. *Acta Crystallogr D Biol Crystallogr* **62**, 32–47.

Garman, E. F., and Schneider, T. R. (1997): Macromolecular Cryocrystallography. *J Appl Crystallogr* **30**, 211–237.

Gavira, J. A., and Garcia-Ruiz, J. M. (2002): Agarose as crystallisation media for proteins II: trapping of gel fibres into the crystals. *Acta Crystallogr D Biol Crystallogr* **58**, 1653–6.

Gavira, J. A., Toh, D., Lopez-Jaramillo, J., Garcia-Ruiz, J. M., and Ng, J. D. (2002): Ab initio crystallographic structure determination of insulin from protein to electron density without crystal handling. *Acta Crystallogr D Biol Crystallogr* **58**, 1147–54.

Georgalis, Y., and Saenger, W. (1993): Time-Resolved Light-Scattering-Studies on Protein Precrystallization Fractal Clusters. *Adv Colloid Interface Sci* **46**, 165–183.

Giacovazzo, C. (ed.) (2002): *Fundamentals of crystallography*. Oxford University Press, Oxford, UK.

Giege, R., Dock, A. C., Kern, D., Lorber, B., Thierry, J. C., and Moras, D. (1986): The Role of Purification in the Crystallization of Proteins and Nucleic-Acids. *J Cryst Growth* **76**, 554–561.

Gilboa, A. J. K., Myles, D. A. A., Habash, J., Raftery, J., and Helliwell, J. R. (2001): Neutron Laue diffraction experiments on a large unit cell: concanavalin A complexed with methyl-alpha-D-glucopyranoside. *J Appl Crystallogr* **34**, 454–457.

Gilliland, G. L., Tung, M., and Ladner, J. (1996): The biological macromolecule crystallization database and NASA protein crystal growth archive. *J Res Nat Inst Stand Technol* **101**, 309–320.

Gilliland, G. L., Tung, M., and Ladner, J. E. (2002): The biological macromolecule crystallization database: crystallization procedures and strategies. *Acta Crystallogr D Biol Crystallogr* **58**, 916–920.

Glaeser, R., Downing, K., DeRosier, D., Chiu, W., and Frank, J. (2007): *Electron crystallography of biological macromolecules* Oxford University Press, Oxford, UK.

Glaeser, R., Facciotti, M., Walian, P., Rouhani, S., Holton, J., MacDowell, A., Celestre, R., Cambie, D., and Padmore, H. (2000): Characterization of conditions required for X-ray diffraction experiments with protein microcrystals. *Biophys J* **78**, 3178–3185.

Glover, I. D., Harris, G. W., Helliwell, J. R., and Moss, D. S. (1991): The variety of X-ray diffuse scattering from macromolecular crystals and its respective components. *Acta Crystallogr B* **47**, 960–968.

Gonzalez, A. (2007): A comparison of SAD and two-wavelength MAD phasing for radiation-damaged Se-MET crystals. *J Synch Radiat* **14**, 43–50.

Gonzalez, A., Cohen, A., Eriksson, T., McPhillips, S., Moorhead, P., Narevicius, J., Sharp, K., Smith, C., Song, J., and Soltis, S. M. (2006): Facility Updates: Remote Access to the SSRL Macromolecular Crystallography Beamlines. *Synch Radiat News* **18**, 36–39.

Gonzalez, A., Denny, R., and Nave, C. (1994): Data collection at short wavelengths in protein crystallography. *Acta Crystallogr D Biol Crystallogr* **50**, 276–82.

Gonzalez, A., Moorhead, P., McPhillips, S. E., Song, J., Sharp, K., Taylor, J. R., Adams, P. D., Sauter, N. K., and Soltis, S. M. (2008): Web-Ice: integrated data collection and analysis for macromolecular crystallography. *J Appl Crystallogr* **41**, 176–184.

Gonzalez, A., and Nave, C. (1994): Radiation-Damage in Protein Crystals at Low-Temperature. *Acta Crystallogr D Biol Crystallogr* **50**, 874–877.

Gonzalez, A., Thompson, A. W., and Nave, C. (1992): Cryo-Protection of Protein Crystals in Intense X-Ray-Beams. *Review of Scientific Instruments* **63**, 1177–1180.

Govada, L., Carpenter, L., da Fonseca, P. C., Helliwell, J. R., Rizkallah, P., Flashman, E., Chayen, N. E., Redwood, C., and Squire, J. M. (2008): Crystal structure of the C1 domain of cardiac myosin binding protein-C: implications for hypertrophic cardiomyopathy. *J Mol Biol* **378**, 387–97.

Govada, L., and Chayen, N. E. (2009): Crystallization by Controlled Evaporation Leading to High Resolution Crystals of the C1 Domain of Cardiac Myosin Binding Protein-C (cMyBP-C). *Cryst Growth & Design* **9**, 1729–1732.

Grabolle, M., Haumann, M., Muller, C., Liebisch, P., and Dau, H. (2006): Rapid loss of structural motifs in the manganese complex of oxygenic photosynthesis by X-ray irradiation at 10-300 K. *J Biol Chem* **281**, 4580–4588.

Gray, H. B., Malmstrom, B. G., and Williams, R. J. P. (2000): Copper coordination in blue proteins. *J Biological Inorganic Chemistry* **5**, 551–559.

Greenhough, T. J., and Helliwell, J. R. (1982a): Oscillation Camera Data-Processing – Reflecting Range and Prediction of Partiality .2. Monochromatized Synchrotron X-Radiation from a Singly Bent Triangular Monochromator. *J Appl Crystallogr* **15**, 493–508.

Greenhough, T. J., and Helliwell, J. R. (1982b): Oscillation camera data processing: reflecting range and prediction of partiality. I. Conventional X-ray sources. *J Appl Crystallogr* **15**, 338–351.

Grigorieff, N., and Henderson, R. (1995): Diffuse scattering in electron diffraction data from protein crystals. *Ultramicroscopy* **60**, 295–309.

Grisshammer, R., and Buchanan, S. K. (2006): *Structural biology of membrane proteins*, Royal Society of Chemistry, Cambridge, UK.

Gulyaeva-Tcherkassova, J., Labiuk, S. L., Lee, J. S., Waygood, E. B., and Delbaere, L. T. (2003): Jel44 monoclonal Fab fragment specific for HPr of the phosphoenolpyruvate:sugar phosphotransferase system of Escherichia coli and the complex of Jel44 Fab fragment with HPr: preparation, crystallization and preliminary crystallographic analysis. *Acta Crystallogr D Biol Crystallogr* **59**, 161–2.

Haas, D. J., and Rossmann, M. G. (1970): Crystallographic studies on lactate dehydrogenase at − 75 °C. *Acta Crystallogr* **B26**, 998–1004.

Hahn, T., Janovec, V., and Klapper, H. (1999): Bicrystals, twins and domain structures – A comparison. *Ferroelectrics* **222**, 269–279.

Hahn, T., and Klapper, H. (2003). In A. Authier (ed.): *International tables for crystallography*. Vol. D, Section 3.3, Kluwer Academic Publishers, Dordrecht.

Haire, L. (1999), pp. 125–128. In T. M. Bergfors (ed.): *Protein crystallization techniques,strategies and tips*, International University Line, California.

Haire Lloyd, L. F. (1996): *Strategies for protein crystal growth- screening and optimisation*, University of London, London.

Hansen, C., and Quake, S. R. (2003): Microfluidics in structural biology: smaller, faster . . . better. *Curr Opin Struct Biol* **13**, 538–544.

Hansen, C. L., Skordalakes, E., Berger, J. M., and Quake, S. R. (2002): A robust and scalable microfluidic metering method that allows protein crystal growth by free interface diffusion. *Proc Natl Acad Sci U S A* **99**, 16531–6.

Hanson, B. L., Harp, J. M., and Bunick, G. J. (2003a): The well-tempered protein crystal: annealing macromolecular crystals. *Methods Enzymol* **368**, 217–35.

Hanson, B. L., Harp, J. M., Kirschbaum, K., Schall, C. A., DeWitt, K., Howard, A., Pinkerton, A. A., and Bunick, G. J. (2002): Experiments testing the abatement of radiation damage in D-xylose isomerase crystals with cryogenic helium. *J Synch Radiat* **9**, 375–81.

Hanson, B. L., Schall, C. A., and Bunick, G. J. (2003b): New techniques in macromolecular cryocrystallography: macromolecular crystal annealing and cryogenic helium. *J Struct Biol* **142**, 77–87.

Harp, J. M., Hanson, B. L., Timm, D. E., and Bunick, G. J. (1999): Macromolecular crystal annealing: evaluation of techniques and variables. *Acta Crystallogr D Biol Crystallogr* **55 (Pt 7)**, 1329–34.

Harp, J. M., Timm, D. E., and Bunick, G. J. (1998): Macromolecular crystal annealing: overcoming increased mosaicity associated with cryocrystallography. *Acta Crystallogr D Biol Crystallogr* **54**, 622–8.

Harris, K. D. M., Tremayne, M., Lightfoot, P., and Bruce, P. G. (1994): Crystal-Structure Determination from Powder Diffraction Data by Monte-Carlo Methods. *J Am Chem Soc* **116**, 3543–3547.

Hauptman, H. (1997): Phasing methods for protein crystallography. *Curr Opin Struct Biol* **7**, 672–680.

Hazemann, I., Dauvergne, M. T., Blakeley, M. P., Meilleur, F., Haertlein, M., Van Dorsselaer, A., Mitschler, A., Myles, D. A., and Podjarny, A. (2005): High-resolution neutron protein crystallography with radically small crystal volumes: application of perdeuteration to human aldose reductase. *Acta Crystallogr D Biol Crystallogr* **61**, 1413–7.

Hedman, B., Hodgson, K. O., Helliwell, J. R., Liddington, R., and Papiz, M. Z. (1985): Protein microcrystal diffraction and the effects of radiation damage with ultra-high-flux synchrotron radiation. *Proc Natl Acad Sci U S A* **82**, 7604–7.

Heijna, M. C. R., Poodt, P. W. G., Tsukamoto, K., de Grip, W. J., Christianen, P. C. M., Maan, J. C., Hendrix, J. L. A., Van Enckevort, W. J. P., and Vlieg, E. (2007): Magnetically controlled gravity for protein crystal growth. *Applied Physics Letters* **90**, 264105.

Helliwell, J. R. (1984): Synchrotron X-Radiation Protein Crystallography – Instrumentation, Methods and Applications. *Reports on Progress in Physics* **47**, 1403–1497.

Helliwell, J. R. (1988): Protein crystal perfection and the nature of radiation damage. *J. Cryst Growth* **90**, 259–272.

Helliwell, J. R. (1991): Macromolecular crystallography using Synchrotron Radiation: Progress with station 9.5 and a novel, toastrack detector scheme for Laue diffraction. *Nuclear Instruments & Methods in Physics Research Section a-Accelerators Spectrometers Detectors and Associated Equipment* **308**, 260–266.

Helliwell, J. R. (1992): *Macromolecular crystallography with synchrotron radiation.* Cambridge University Press, Cambridge, UK. Paperback version (2005).

Helliwell, J. R. (2005): Protein crystal perfection and its application. *Acta Crystallogr D Biol Crystallogr* **61**, 793–798.

Helliwell, J. R. (2008): Macromolecular crystal twinning, lattice disorders and multiple crystals *Crystallogr Rev* **14**, 189–250.

Helliwell, J. R., and Fourme, R. (1983): The ESRF as a facility for protein crystallography: A report and design study, pp. 1–36: *ESRP Report.* Geneva (available from ESRF, Grenoble, France).

Helliwell, J. R., Glover, I. D., Jones, A., Pantos, E., and Moss, D. S. (1986a): Protein Dynamics – Use of Computer-Graphics and Protein Crystal Diffuse-Scattering Recorded with Synchrotron X-Radiation. *Biochemical Society Transactions* **14**, 653–655.

Helliwell, J. R., Greenhough, T. J., Carr, P. D., Rule, S. A., Moore, P. R., Thompson, A. W., and Worgan, J. S. (1982): Central Data-Collection Facility for Protein

Crystallography, Small-Angle Diffraction and Scattering at the Daresbury-Laboratory Synchrotron Radiation Source (SRS), England. *J Phys E-Sci Instrum* **15**, 1363–1372.

Helliwell, J. R., Habash, J., Cruickshank, D. W. J., Harding, M. M., Greenhough, T. J., Campbell, J. W., Clifton, I. J., Elder, M., Machin, P. A., Papiz, M. Z., and Zurek, S. (1989): The Recording and Analysis of Synchrotron X-Radiation Laue Diffraction Photographs. *J Appl Crystallogr* **22**, 483–497.

Helliwell, J. R., Helliwell, M., and Jones, R. H. (2005): Ab initio structure determination using dispersive differences from multiple-wavelength synchrotron-radiation powder diffraction data. *Acta Crystallogr Section A* **61**, 568–574.

Helliwell, J. R., Papiz, M. Z., Glover, I. D., Habash, J., Thompson, A. W., Moore, P. R., Harris, N., Croft, D., and Pantos, E. (1986b): The Wiggler Protein Crystallography Workstation at the Daresbury SRS – Progress and Results. *Nucl Instrum Methods Phys Res Section a-Accel Spectrom Detect AssoEquip* **246**, 617–623.

Helliwell, J. R., Snell, E., and Weisgerber, S. (1996): An investigation of the perfection of lysozyme crystals grown in microgravity and on earth, pp. 155–170. In L. Ratke, H. Walter, and B. Feuerbache (ed.): *Lecture notes in physics*, Springer-Verlag, Heidelberg.

Henderson, R. (1990): Cryo-protection of protein crystals against radiation damage in electron and X-ray diffraction. *Proc. R. Soc. Lond. B.* **241**, 6–8.

Henderson, R., and Unwin, N. (1976): 3-Dimensional Structure of Purple Membrane. *Biophys J* **16**, A1–A1.

Henderson, R., and Unwin, P. N. T. (1975): 3-Dimensional Model of Purple Membrane Obtained by Electron-Microscopy. *Nature* **257**, 28–32.

Hendrickson, W. A. (1976): Radiation-Damage in Protein Crystallography. *J Molec Biol* **106**, 889–893.

Hendrickson, W. A., Horton, J. R., and Lemaster, D. M. (1990): Selenomethionyl Proteins Produced for Analysis by Multiwavelength Anomalous Diffraction (MAD) – a Vehicle for Direct Determination of 3-Dimensional Structure. *Embo Journal* **9**, 1665–1672.

Henkamer, B., Chayen,N.E., De Las Rivias, J. & Barber, J. (1992): Photosystem II cor complexes and CP-43 crystalliztion trials. First Robert Hill Symposium on Photsynthesis, pp. 11–12.

Henrich, S., Cameron, A., Bourenkov, G. P., Kiefersauer, R., Huber, R., Lindberg, I., Bode, W., and Than, M. E. (2003): The crystal structure of the proprotein processing proteinase furin explains its stringent specificity. *Nature Struct Biol* **10**, 669–669.

Heras, B., and Martin, J. L. (2005): Post-crystallization treatments for improving diffraction quality of protein crystals. *Acta Crystallogr D Biol Crystallogr* **61**, 1173–1180.

Herbst-Irmer, R., and Sheldrick, G. M. (2002): Refinement of obverse/reverse twins. *Acta Crystallogr B* **58**, 477–81.

Hery, S., Genest, D., and Smith, J. C. (1998): X-ray diffuse scattering and rigid-body motion in crystalline lysozyme probed by molecular dynamics simulation. *J Molec Biol* **279**, 303–319.

Hirvonen, M., and Papageorgiou, A. C. (2002): Crystallization and preliminary crystallographic analysis of a family 45 endoglucanase from the thermophilic fungus Melanocarpus albomyces. *Acta Crystallogr D Biol Crystallogr* **58**, 336–8.

Hodgkin, D. C., Kamper, J., Lindsay, J., MacKay, M., Pickworth, J., Robertson, J.H., Shoemaker, C. B., White, J. G., Prosen, R. J., and Trueblood, K. N. (1957): The Structure of Vitamin B12 I. An Outline of the Crystallographic Investigation of Vitamin B12. *Proc. R. Soc. Lond. A.* **242**, 228–263.

Hope, H. (1988): Cryocrystallography of biological macromolecules: a generally applicable method. *Acta Crystallogr B* **44**, 22–6.

Hope, H. (1990): Crystallography of biological macromolecules at ultra-low temperature. *Annu Rev Biophys Biophys Chem* **19**, 107–26.

Hope, H., Frolow, F., von Bohlen, K., Makowski, I., Kratky, C., Halfon, Y., Danz, H., Webster, P., Bartels, K. S., Wittmann, H. G., and *et al.* (1989): Cryocrystallography of ribosomal particles. *Acta Crystallogr B* **45**, 190–9.

Horsefield, R., Yankovskaya, V., Tornroth, S., Luna-Chavez, C., Stambouli, E., Barber, J., Byrne, B., Cecchini, G., and Iwata, S. (2003): Using rational screening and electron microscopy to optimize the crystallization of succinate:ubiquinone oxidoreductase from Escherichia coli. *Acta Crystallogr D Biol Crystallogr* **59**, 600–2.

Houvee-Levin, C., and Sicard-Roselli, C. (2001), pp. 553–584. In C. D. Jonah, and B. S. M. Rao (ed.): *Studies in Physical and theoretical chemistry no. 87 – radiation chemistry*, Elsevier, Amsterdam.

Hu, Z. W., Thomas, B. R., and Chernov, A. A. (2001): Laboratory multiple-crystal X-ray topography and reciprocal-space mapping of protein crystals: influence of impurities on crystal perfection. *Acta Crystallogr D Biol Crystallogr* **57**, 840–846.

Hünefeld, F. L. (1840): *Der Chemismus in der thierischen Organisation*, 158–163.

Hunte, C., Koepke, J., Lange, C., Rossmanith, T., and Michel, H. (2000): Structure at 2.3 A resolution of the cytochrome bc(1) complex from the yeast Saccharomyces cerevisiae co-crystallized with an antibody Fv fragment. *Structure* **8**, 669–84.

Hunte, C., and Michel, H. (2002): Crystallisation of membrane proteins mediated by antibody fragments. *Curr Opin Struct Biol* **12**, 503–8.

Hunte, C., von Jagow, G., and Schagger, H. (2002): *Membrane protein purification and crystallization: A practical guide*, Academic Press, San Diego, USA.

Huxley, H. E., and Kendrew, J. C. (1953): Discontinious lattice changes in Haemoglobin Crystals. *Acta Crystallogr* **6**, 76–80.

Igarashi, N., Ikuta, K., Miyoshi, T., Matsugaki, N., Yamada, Y., Yousef, M. S., and Wakatsuki, S. (2008): X-ray beam stabilization at BL-17A, the protein microcrystallography beamline of the Photon Factory. *J Synch Radiat* **15**, 292–295.

Ihee, H., Rajagopal, S., Srajer, V., Pahl, R., Anderson, S., Schmidt, M., Schotte, F., Anfinrud, P. A., Wulff, M., and Moffat, K. (2005): Visualizing reaction pathways in photoactive yellow protein from nanoseconds to seconds. *Proc Nat Acad Sci USA* **102**, 7145–7150.

Islam, S. A., Carvin, D., Sternberg, M. J. E., and Blundell, T. L. (1998): HAD, a data bank of heavy-atom binding sites in protein crystals: a resource for use in multiple isomorphous replacement and anomalous scattering. *Acta Crystallogr D Biol Crystallogr* **54**, 1199–1206.

Iwasaki, W., Sasaki, H., Nakamura, A., Kohama, K., and Tanokura, M. (1999): Crystallization and preliminary X-ray diffraction studies of a 40 kDa calcium binding protein specifically expressed in plasmodia of Physarum polycephalum. *J Biochemistry* **126**, 7–9.

Iwata, S. (2003): *Methods and results in crystallization of membrane proteins*, International University Line.

Jaakola, V. P., Griffith, M. T., Hanson, M. A., Cherezov, V., Chien, E. Y. T., Lane, J. R., IJzerman, A. P., and Stevens, R. C. (2008): The 2.6 Angstrom Crystal Structure of a Human A(2A) Adenosine Receptor Bound to an Antagonist. *Science* **322**, 1211–1217.

Jakoncic, J., Di Michiel, M., Zhong, Z., Honkimaki, V., Jouanneau, Y., and Stojanoff, V. (2006): Anomalous diffraction at ultra-high energy for protein crystallography. *J Appl Crystallogr* **39**, 831–841.

Jancarik, J., and Kim, S.-H. (1991): Sparse matrix sampling: a screening method for crystallization of proteins. *J Appl Crystallogr* **24**, 409–411.

Jenner, M. J., Wright, J. P., Margiolaki, I., and Fitch, A. N. (2007): Successful protein cryocooling for powder diffraction. *J Appl Crystallogr* **40**, 121–124.

Johansson, L. C., Wohri, A. B., Katona, G., Engstrom, S., and Neutze, R. (2009): Membrane protein crystallization from lipidic phases. *Curr Opin Struct Biol*, **19**, 372–378.

Jovine, L., Djordjevic, S., and Rhodes, D. (2000): The crystal structure of yeast phenylalanine tRNA at 2.0 A resolution: cleavage by Mg(2+) in 15-year old crystals. *J Mol Biol* **301**, 401–14.

Juers, D. H., Lovelace, J., Bellamy, H. D., Snell, E. H., Matthews, B. W., and Borgstahl, G. E. (2007): Changes to crystals of Escherichia coli beta-galactosidase during room-temperature/low-temperature cycling and their relation to cryo-annealing. *Acta Cryst. D* **63**, 1139–1153.

Juers, D. H., and Matthews, B. W. (2004a): Cryo-cooling in macromolecular crystallography: advantages, disadvantages and optimization. *Quart Rev Biophys* **37**, 105–119.

Juers, D. H., and Matthews, B. W. (2004b): The role of solvent transport in cryo-annealing of macromolecular crystals. *Acta Crystallogr D Biol Crystallogr* **60**, 412–421.

Kabsch, W. (1998): Automatic indexing of rotation diffraction patterns. *J Appl Crystallogr* **21**, 67–72.

Kalinin, Y., Kmetko, J., Bartnik, A., Stewart, A., Gillilan, R., Lobkovsky, E., and Thorne, R. (2005): A new sample mounting technique for room-temperature macromolecular crystallography. *J Appl Crystallogr* **38**, 333–339.

Kam, Z., Shore, H. B., and Feher, G. (1978): Crystallization of Proteins. *J Molec Biol* **123**, 539–555.

Kantardjieff, K. A., and Rupp, B. (2003): Matthews coefficient probabilities: Improved estimates for unit cell contents of proteins, DNA, and protein-nucleic acid complex crystals. *Protein Sci* **12**, 1865–1871.

Kariuki, B. M., Serrano-Gonzalez, H., Johnston, R. L., and Harris, K. D. M. (1997): The application of a genetic algorithm for solving crystal structures from powder diffraction data. *Chemical Physics Letters* **280**, 189–195.

Kauffmann, B., Weiss, M. S., Lamzin, V. S., and Schmidt, A. (2006): How to avoid premature decay of your macromolecular crystal: a quick soak for long life. *Structure* **14**, 1099–105.

Kawabata, K., Saitoh, K., Takahashi, M., Sugahara, M., Asama, H., Mishima, T., and Miyano, M. (2006): Integrated state evaluation for the images of crystallization droplets utilizing linear and nonlinear classifiers. *Acta Crystallogr D Biol Crystallogr* **62**, 1066–1072.

Kempkes, R., Stofko, E., Lam, K., and Snell, E. H. (2008): Glycerol concentrations required for the successful vitrification of cocktail conditions in a high-throughput crystallization screen. *Acta Crystallogr D.* **64**, 287–301.

Kerfeld, C. A., Wu, Y. P., Chan, C., Krogmann, D. W., and Yeates, T. O. (1997): Crystals of the carotenoid protein from Arthrospira maxima containing uniformly oriented pigment molecules. *Acta Crystallogr D Biol Crystallogr* **53**, 720–723.

Kiefersauer, R., Than, M. E., Dobbek, H., Gremer, L., Melero, M., Strobl, S., Dias, J. M., Soulimane, T., and Huber, R. (2000): A novel free-mounting system for protein crystals: transformation and improvement of diffraction power by accurately controlled humidity changes. *J Appl Crystallogr* **33**, 1223–1230.

Kim, T. K., Zuo, X. B., Tiede, D. M., and Ihee, H. (2004): Exploring fine structures of photoactive yellow protein in solution using wide-angle X-ray scattering. *Bulletin of the Korean Chemical Society* **25**, 1676–1680.

Kim, Y., Quartey, P., Li, H., Volkart, L., Hatzos, C., Chang, C., Nocek, B., Cuff, M., Osipiuk, J., Tan, K., Fan, Y., Bigelow, L., Maltseva, N., Wu, R., Borovilos, M., Duggan, E., Zhou, M., Binkowski, T. A., Zhang, R. G., and Joachimiak, A. (2008): Large-scale evaluation of protein reductive methylation for improving protein crystallization. *Nature Methods* **5**, 853–4.

King, M. V. (1958): Improved Resolution of X-ray Diffraction Patterns of Protein Crystals at Low Temperature. *Nature* **25**, 263–264.

King, M. V. (1965): Macroscopic characteristics of ribonuclease crystals of modifications I and II. *Acta Crystallogr* **18**, 686–689.

Kingston, R. L., Baker, H. M., and Baker, E. N. (1994): Search Designs for Protein Crystallization Based on Orthogonal Arrays. *Acta Crystallogr D Biol Crystallogr* **50**, 429–440.

Koch, M., Breithaupt, C., Kiefersauer, R., Freigang, J., Huber, R., and Messerschmidt, A. (2004): Crystal structure of protoporphyrinogen IX oxidase: a key enzyme in haem and chlorophyll biosynthesis. *Embo Journal* **23**, 1720–1728.

Komatsu, H., Miyashita, S., and Suzuki, Y. (1993): Interferometric Observation of the Interfacial Concentration Gradient Layers around a Lysozyme Crystal. *Jpn J Appl Phys Part 2-Lett* **32**, L1855–L1857.

Korkhin, Y. M., Evdokimov, A., and Shaw Stewart, P. D. (1995): Crystallisation of a protein by microseeding after establishing its phase diagram., Application Note I Douglas Instruments.

Koszelak-Rosenblum, M., Krol, A., Mozumdar, N., Wunsch, K., Ferrin, A., Cook, E., Veatch, C. K., Nagel, R., Luft, J. R., DeTitta, G. T., and Malkowski, M. G. (2009): Determination and application of empirically derived detergent phase boundaries to effectively crystallize membrane proteins *Protein Sci* **18**, 1828–1839.

Koszelak, S., Day, J., Leja, C., Cudney, R., and McPherson, A. (1995): Protein and virus growth on international microgravity laboratory-2. *Biophys J* **69**, 13–19.

Koth, C. M., Orlicky, S. M., Larson, S. M., and Edwards, A. M. (2003): Use of limited proteolysis to identify protein domains suitable for structural analysis. *Methods Enzymol* **368**, 77–84.

Kouyama, T., Nishikawa, T., Tokuhisa, T., and Okumura, H. (2004): Crystal structure of the L intermediate of bacteriorhodopsin: Evidence for vertical translocation of a water molecule during the proton pumping cycle. *J Molec Biol* **335**, 531–546.

Kovari, L. C., Momany, C., and Rossmann, M. G. (1995): The use of antibody fragments for crystallization and structure determinations. *Structure* **3**, 1291–1293.

Krengel, U., Dey, R., Sasso, S., Okvist, M., Ramakrishnan, C., and Kast, P. (2006): Preliminary X-ray crystallographic analysis of the secreted chorismate mutase from Mycobacterium tuberculosis: a tricky crystallization problem solved. *Acta Crystallogr Section F-Struct Biol Crystalliz Commun* **62**, 441–445.

Kriminski, S., Caylor, C. L., Nonato, M. C., Finkelstein, K. D., and Thorne, R. E. (2002): Flash-cooling and annealing of protein crystals. *Acta Crystallogr D Biol Crystallogr* **58**, 459–471.

Kuhn, P., Wilson, K., Patch, M. G., and Stevens, R. C. (2002): The genesis of high-throughput structure-based drug discovery using protein crystallography. *Curr Opin Chem Biol* **6**, 704–10.

Kurihara, K., Miyashita, S., Sazaki, G., Nakada, T., Suzuki, Y., and Komatsu, H. (1996): Interferometric study on the crystal growth of tetragonal lysozyme crystal. *J Cryst Growth* **166**, 904–908.

Kuznetsov, Y. G., Malkin, A. J., Greenwood, A., and McPherson, A. (1995): Interferometric Studies of Growth-Kinetics and Surface-Morphology in Macromolecular Crystal-Growth – Canavalin, Thaumatin, and Turnip Yellow Mosaic-Virus. *J Struct Biol* **114**, 184–196.

Kyrieleis, O. J. P., Goettig, P., Kiefersauer, R., Huber, R., and Brandstetter, H. (2005): Crystal structures of the tricorn interacting factor F3 from Thermoplasma acidophilum, a zinc aminopeptidase in three different conformations. *J Molec Biol* **349**, 787–800.

Kyte, J., and Doolittle, R. F. (1982): A Simple Method for Displaying the Hydropathic Character of a Protein. *J Molec Biol* **157**, 105–132.

Lagartera, L., Gonzalez, A., Hermoso, J. A., Saiz, J. L., Garcia, P., Garcia, G. L., and Menendez, M. (2005): Pneumococcal phosphorylcholine esterase, Pce, contains a metal binuclear center that is essential for substrate binding and catalysis. *Protein Sci* **14**, 3013–3024.

Landau, E. M., and Rosenbusch, J. P. (1996): Lipidic cubic phases: A novel concept for the crystallization of membrane proteins. *Proc Nat Acad Sci USA* **93**, 14532–14535.

Langan, P., and Greene, G. (2004): Protein crystallography with spallation neutrons: collecting and processing wavelength-resolved Laue protein data. *J Appl Crystallogr* **37**, 253–257.

Langan, P., Greene, G., and Schoenborn, B. P. (2004): Protein crystallography with spallation neutrons: the user facility at Los Alamos Neutron Science Center. *J Appl Crystallogr* **37**, 24–31.

Larson, S. B., Day, J. S., Cudney, R., and McPherson, A. (2007): A novel strategy for the crystallization of proteins: X-ray diffraction validation. *Acta Crystallogr D Biol Crystallogr* **63**, 310–8.

Lattman, E. E. (2001): Molecular structures from femtosecond X-ray pulses. *Proc Natl Acad Sci U S A* **98**, 6535–6.

Lau, B. T. C., Baitz, C. A., Dong, X. P., and Hansen, C. L. (2007): A complete microfluidic screening platform for rational protein crystallization. *J Am Chem Soc* **129**, 454–455.

Lebedev, A. A., Vagin, A. A., and Murshudov, G. N. (2006): Intensity statistics in twinned crystals with examples from the PDB. *Acta Crystallogr D Biol Crystallogr* **62**, 83–95.

Leslie, A. G. W. (1992): Recent changes to the MOSFLM package for processing film and image plate data. *Joint CCP4 + ESF-EAMCB Newsletter on Protein Crystallography*. STFC, Daresbury, UK

Lewis, R. (1994): Multiwire Gas Proportional Counters: Decrepit Antiques or Classic Performers? *J. Synch Rad.* **1**, 43–53.

Lietzke, S. E., Carperos, V. E., and Kundrot, C. E. (1996): Data reduction from twinned RNA crystals. *Acta Crystallogr D Biol Crystallogr* **52**, 687–692.

Lin, S. X., Zhou, M., Azzi, A., Xu, G. J., Wakayama, N. I., and Ataka, M. (2000): Magnet used for protein crystallization: novel attempts to improve the crystal quality. *Biochem Biophys Res Commun* **275**, 274–8.

Loll, P. J., Allaman, M., and Wiencek, J. (2001): Assessing the role of detergent-detergent interactions in membrane protein crystallization. *J Cryst Growth* **232**, 432–438.

Longnecker, K. L., Garrard, S. M., Sheffield, P. J., and Derewenda, Z. S. (2001): Protein crystallization by rational mutagenesis of surface residues: Lys to Ala mutations promote crystallization of RhoGDI. *Acta Crystallogr D Biol Crystallogr* **57**, 679–688.

Lott, J. S., Banfield, M. J., Sigrell, J., and Baker, E. N. (2003): Making the most of two crystals: structural analysis of a conserved hypothetical protein using native gel screening and SAD phasing. *Acta Crystallogr D Biol Crystallogr* **59**, 2242–2246.

Lovelace, J., Snell, E. H., Pokross, M., Arvai, A. S., Nielsen, C., Xuong, N.-H., Bellamy, H. D., and Borgstahl, G. E. O. (2000): BEAM-ish: a graphical user interface for the physical characterization of macromolecular crystals. *J Appl Crystallogr* **33**, 1187–1188.

Lovelace, J. J., Murphy, C. R., Daniels, L., Narayan, K., Schutt, C. E., Lindberg, U., Svensson, C., and Borgstahl, G. E. O. (2008): Protein crystals can be incommensurately modulated. *J Appl Crystallogr* **41**, 600–605.

Lovelace, J. J., Soares, A. S., Bellamy, H. D., Sweet, R. M., Snell, E. H., and Borgstahl, G. E. (2004): First results of digital topography applied to macromolecular crystals. *J Appl Crystallogr* **37**, 481–485.

Low, B. W., and Richards, F. M. (1954): Measurements of the Density, Composition and Related Unit Cell Dimensions of Some Protein Crystals. *J the Americal Chemical Society* **76**, 2511–2518.

Lu, Z. P., and Liu, C. T. (2003): Glass formation criterion for various glass-forming systems. *Phys. Rev. Lett.* **91**, 115505, 1–4.

Luft, J. R., Collins, R. J., Fehrman, N. A., Lauricella, A. M., Veatch, C. K., and DeTitta, G. T. (2003): A deliberate approach to screening for initial crystallization conditions of biological macromolecules. *J Struct Biol* **142**, 170–9.

Luft, J. R., and DeTitta, G. T. (1999): A method to produce microseed stock for use in the crystallization of biological macromolecules. *Acta Crystallogr D Biol Crystallogr* **55**, 988–993.

Luft, J. R., Wolfley, J., Jurisica, I., Glasgow, J., Fortier, S., and DeTitta, G. T. (2001): Macromolecular crystallization in a high throughput laboratory-the search phase. *J Cryst Growth* **232**, 591–595.

Luft, J. R., Wolfley, J. R., Said, M. I., Nagel, R. M., Lauricella, A. M., Smith, J. L., Thayer, M. H., Veatch, C. K., Snell, E. H., Malkowski, M. G., and DeTitta, G. T. (2007): The efficient optimization of crystallization conditions – Successful application of drop volume ration and temperature screening. *Protein Sci* **16**, 715–722.

Lundblad, R. L. (2005): *Chemical reagents for protein modification*, CRC Press, Boca Raton.

Lunde, C. S., Rouhani, S., Remis, J. P., and Glaeser, R. M. (2008): Microcrystal screening with a novel design for beamline-mountable crystallization wells. *J Appl Crystallogr* **41**, 483–486.

Lunde, C. S., Rouhani, S., Remis, J. P., Ruzin, S. E., Ernst, J. A., and Glaeser, R. M. (2005): UV microscopy at 280 nm is effective in screening for the growth of protein microcrystals. *J Appl Crystallogr* **38**, 1031–1034.

Lunin, V. Y., Lunina, N. L., and Baumstark, M. W. (2007): Estimates of the twinning fraction for macromolecular crystals using statistical models accounting for experimental errors. *Acta Crystallogr D Biol Crystallogr* **63**, 1129–1138.

Mac Sweeney, A., and D'Arcy, A. (2003): A simple and rapid method for mounting protein crystals at room temperature. *J Appl Crystallogr* **36**, 165–166.

MacRae, I. J., and Doudna, J. A. (2007): An unusual case of pseudo-merohedral twinning in orthorhombic crystals of Dicer. *Acta Crystallogr D Biol Crystallogr* **63**, 993–999.

MacRae, I. J., Zhou, K. H., Li, F., Repic, A., Brooks, A. N., Cande, W. Z., Adams, P. D., and Doudna, J. A. (2006): Structural basis for double-stranded RNA processing by dicer. *Science* **311**, 195–198.

Maeda, M., Chatake, T., Tanaka, I., Ostermann, A., and Niimura, N. (2004): Crystallization of a large single crystal of cubic insulin for neutron protein crystallography. *J Synch Radiat* **11**, 41–44.

Makino, D. L., Henschen-Edman, A. H., Larson, S. B., and McPherson, A. (2007): Bence–Jones KWR protein structures determined by X-ray crystallography. *Acta Crystallogr D Biol Crystallogr* **63**, 780–92.

Makino, D. L., Henschen-Edman, A. H., and McPherson, A. (2005): Four crystal forms of a Bence–Jones protein. *Acta Crystallogr Sect F Struct Biol Cryst Commun* **61**, 79–82.

Malkin, A. J., Cheung, J., and McPherson, A. (1993): Crystallization of Satellite Tobacco Mosaic-Virus .1. Nucleation Phenomena. *J Cryst Growth* **126**, 544–554.

Malkin, A. J., and McPherson, A. (1993): Light-Scattering Investigations of Protein and Virus Crystal-Growth – Ferritin, Apoferritin and Satellite Tobacco Mosaic-Virus. *J Cryst Growth* **128**, 1232–1235.

Mancini, E. J., Grimes, J. M., Malby, R., Sutton, G. C., Kainov, D. E., Juuti, J. T., Makeyev, E. V., Tuma, R., Bamford, D. H., and Stuart, D. I. (2003): Order and disorder in crystals of hexameric NTPases from dsRNA bacteriophages. *Acta Crystallogr D Biol Crystallogr* **59**, 2337–2341.

Mao, X., and Chen, X. (2005): Crystallization and X-ray crystallographic analysis of human STAT1. *Acta Crystallogr Sect F Struct Biol Cryst Commun* **61**, 666–8.

Marchesini, S., Boutet, S., Sakdinawat, A. E., Bogan, M. J., Bajt, S., Barty, A., Chapman, H. N., Frank, M., Hau-Riege, S. P., Szöke, A., Congwu Cui, Shapiro, D. A., Howells, M. R., Spence, J. C. H., Shaevitz, J. W., Lee, J. Y., Hajdu, J., and Seibert, M. M. (2008): Ultrafast, ultrabright, X-ray holography using a uniformly redundant array. *Nature Photon* **2**, 560–563.

Margiolaki, I., and Wright, J. P. (2008): Powder crystallography on macromolecules. *Acta Crystallogr A* **64**, 169–80.

Margiolaki, I., Wright, J. P., Fitch, A. N., Fox, G. C., Labrador, A., Von Dreele, R. B., Miura, K., Gozzo, F., Schiltz, M., Besnard, C., Camus, F., Pattison, P., Beckers, D., and Degen, T. (2007a): Powder diffraction studies on proteins: An overview of data collection approaches. *Z. Kristallogr. Suppl.* **26**, 1–13.

Margiolaki, I., Wright, J. P., Fitch, A. N., Fox, G. C., and von Dreele, R. B. (2005): Synchrotron X-ray powder diffraction study of hexagonal turkey egg-white lysozyme. *Acta Crystallogr D Biol Crystallogr* **61**, 423–432.

Margiolaki, I., Wright, J. P., Wilmanns, M., Fitch, A. N., and Pinotsis, N. (2007b): Second SH3 domain of ponsin solved from powder diffraction. *J Am Chem Soc* **129**, 11865–11871.

Mateja, A., Devedjiev, Y., Krowarsch, D., Longenecker, K., Dauter, Z., Otlewski, J., and Derewenda, Z. S. (2002): The impact of Glu>Ala and Glu>Asp mutations on the crystallization properties of RhoGDI: the structure of RhoGDI at 1.3 A resolution. *Acta Crystallogr D Biol Crystallogr* **58**, 1983–91.

Matsui, Y., Sakai, K., Murakami, M., Shiro, Y., Adachi, S., Okumura, H., and Kouyama, T. (2002): Specific damage induced by X-ray radiation and structural changes in the primary photoreaction of bacteriorhodopsin. *J Mol Biol* **324**, 469–81.

Matthews, B. W. (1968): Solvent content of protein crystals. *J Mol Biol* **33**, 491–7.

Mayo, C. J., Diprose, J. M., Walter, T. S., Berry, I. M., Wilson, J., Owens, R. J., Jones, E. Y., Harlos, K., Stuart, D. I., and Esnouf, R. M. (2005): Benefits of automated crystallization plate tracking, imaging, and analysis. *Structure* **13**, 175–182.

McFerrin, M., and Snell, E. H. (2002): The development and application of a method to quantify the quality of cryoprotectant conditions using standard area detector X-ray images. *J Appl Crystallogr* **35**, 538–545.

McPherson, A. (1991): A brief history of protein crystal growth. *J. Cyst. Growth* **110**, 1–10.

McPherson, A. (1992): Two approaches to the rapid screening of crystallization conditions. *J Cryst Growth* **122**, 161–167.

McPherson, A. (1995): Increasing the size of microcrystals by fine sampling of pH limits. *J Appl Crystallogr* **28**, 362–365.

McPherson, A. (1999): *Crystallization of biological macromolecules*. Cold Spring Harbor Laboratory Press, New York, USA p. 586.

McPherson, A., and Cudney, B. (2006): Searching for silver bullets: an alternative strategy for crystallizing macromolecules. *J Struct Biol* **156**, 387–406.

McPherson, A., Day, J., and Harris, L. J. (2004): Crystals of the beta-subunit of bovine luteinizing hormone and indicators for the involvement of proteolysis in protein crystallization. *Acta Crystallogr D Biol Crystallogr* **60**, 872–877.

McPherson, A., and Giege, R. (2007): Crystallogenesis research for biology in the last two decades as seen from the International Conferences on the Crystallization of Biological Macromolecules. *Cryst Growth & Design* **7**, 2126–2133.

McPherson, A., Greenwood, A., and Day, J. (1991): The effect of microgravity on protein crystal growth. *Adv. Space Res.* **11**, 343–356.

McPherson, A., Malkin, A. J., and Kuznetsov, Y. G. (1995): The Science of Macromolecular Crystallization. *Structure* **3**, 759–768.

McPherson, A., Malkin, A. J., Kuznetsov, Y. G., and Plomp, M. (2001): Atomic force microscopy applications in macromolecular crystallography. *Acta Crystallogr D Biol Crystallogr* **57**, 1053–60.

McPherson, A., Nguyen, C., Larson, S. B., Day, J. S., and Cudney, B. (2007): Development of an alternative approach to protein crystallization. *J Struct Funct Genomics.*

McPhillips, T. M., McPhillips, S. E., Chiu, H. J., Cohen, A. E., Deacon, A. M., Ellis, P. J., Garman, E., Gonzalez, A., Sauter, C., Phizackerley, R. P., Soltis, S. M., and

Kuhn, P. (2002): Blu-Ice and the Distributed Control System: software for data acquisition and instrument control at macromolecular crystallography beamlines. *J. Synch Radiat* **9**, 401–406.

Means, G. E., and Feeney, R. E. (1995): Reductive Alkylation of Proteins. *Anal Biochem* **224**, 1–16.

Merritt, E. A., Sarfaty, S., Feil, I. K., and Hol, W. G. J. (1997): Structural foundation for the design of receptor antagonists targeting Escherichia coli heat-labile enterotoxin. *Structure* **5**, 1485–1499.

Miao, J., Charalambous, P., Kirz, J., and Sayre, D. (1999): Extending the methodology of X-ray crystallography to allow imaging of micrometre-sized non-crystalline specimens. *Nature* **400**, 342–344.

Miao, J., Hodgson, K. O., and Sayre, D. (2001): An approach to three-dimensional structures of biomolecules by using single-molecule diffraction images. *Proc Natl Acad Sci U S A* **98**, 6641–5.

Miao, J., and Sayre, D. (2000): On possible extensions of X-ray crystallography through diffraction- pattern oversampling. *Acta Crystallogr A* **56**, 596–605.

Mikol, V., Hirsch, E., and Giege, R. (1990): Diagnostic of precipitant for biomacromolecule crystallization by quasi-elastic light-scattering. *J Mol Biol* **213**, 187–95.

Mikol, V., Rodeau, J. L., and Giege, R. (1989): Changes of Ph during Biomacromolecule Crystallization by Vapour Diffusion Using Ammonium-Sulfate as the Precipitant. *J Appl Crystallogr* **22**, 155–161.

Miller, M. D., Brinen, L. S., Cohen, A., Deacon, A. M., Ellis, P., McPhillips, S. E., McPhillips, T. M., Phizackerley, R. P., Soltis, S. M., van den Bedem, H., Wolf, G., Xu, Q., and Zhang, Z. (2004): *Synchrotron Radiation Instrumentation: Eighth International Conference on Synchrotron Radiation Instrumentation*, edited by T. Warwick, J. Stöhr, H. A. Padmore & J. Arthur. American Institute of Physics, Melville, New York, USA, pp. 1233–1236.

Mills, D. M., Helliwell, J. R., Kvick, A., Ohta, T., Robinson, I. A., and Authier, A. (2005): Report of the Working Group on Synch Radiat Nomenclature – brightness, spectral brightness or brilliance? *J Synch Radiat* **12**, 385.

Minichino, A., Habash, J., Raftery, J., and Helliwell, J. R. (2003): The properties of (2Fo – Fc) and (Fo – Fc) electron-density maps at medium-to-high resolutions. *Acta Crystallogr D Biol Crystallogr* **59**, 843–9.

Mitchell, E. P., and Garman, E. F. (1994): Flash freezing of protein crystals: investigation of mosaic spread and diffraction limit with variation of cryoprotectant concentration. *J Appl Crystallogr* **27**, 1070–1074.

Miyashita, S., Komatsu, H., Suzuki, Y., and Nakada, T. (1994): Observation of the Concentration Distribution around a Growing Lysozyme Crystal. *J Cryst Growth* **141**, 419–424.

Miyatake, H., Kim, S. H., Motegi, I., Matsuzaki, H., Kitahara, H., Higuchi, A., and Miki, K. (2005): Development of a fully automated macromolecular crystallization/observation robotic system, HTS-80. *Acta Crystallogr D Biol Crystallogr* **61**, 658–663.

Moreno, A., Saridakis, E., and Chayen, N. E. (2002): Combination of oils and gels for enhancing the growth of protein crystals. *J Appl Crystallogr* **35**, 140–142.

Morozov, V. N., and Morozova, T. Y. (1981): Viscoelastic Properties of Protein Crystals – Triclinic Crystals of Hen Egg-White Lysozyme in Different Conditions. *Biopolymers* **20**, 451–467.

Morozov, V. N., Morozova, T. Y., Myachin, E. G., and Kachalova, G. S. (1985): Metastable State of a Protein Crystal. *Acta Crystallogr Section B-Structural Science* **41**, 202–205.

Moss, D. S., Harris, G. W., Wostrack, A., and Sansom, C. (2003): Diffuse X-Ray scattering from molecular crystals *Crystallogr Rev* **9**, 229–277.

Moukhametzianov, R., Burghammer, M., Edwards, P. C., Petitdemange, S., Popov, D., Fransen, M., McMullan, G., Schertler, G. F., and Riekel, C. (2008): Protein crystallography with a micrometre-sized synchrotron-radiation beam. *Acta Crystallogr D Biol Crystallogr* **64**, 158–66.

Murray-Rust, P. (1973): The crystal structure of $[Co(NH_3)_6]_4Cu_5Cl_{17}$: a twinned cubic crystal. *Acta Crystallogr B.* **29**, 2259–2566.

Murray, J., and Garman, E. (2002): Investigation of possible free-radical scavengers and metrics for radiation damage in protein cryocrystallography. *J Synch Radiat* **9**, 347–54.

Murray, J. W., Garman, E. F., and Ravelli, R. B. G. (2004): X-ray absorption by macromolecular crystals: the effects of wavelength and crystal composition on absorbed dose. *J Appl Crystallogr* **37**, 513–522.

Murray, J. W., Rudino-Pinera, E., Owen, R. L., Grininger, M., Ravelli, R. B. G., and Garman, E. F. (2005): Parameters affecting the X-ray dose absorbed by macromolecular crystals. *J Synch Radiat* **12**, 268–275.

Myles, D. A. A. (2006): Neutron protein crystallography: current status and a brighter future. *Curr Opin Struct Biol* **16**, 630–637.

Nagel, R. M., Luft, J. R., and Snell, E. H. (2008): An Aid to Crystallization Optimization – AutoSherlock. *J Appl Crystallogr*, 1173–1176.

Navaza, J. (2008): Combining X-ray and electron-microscopy data to solve crystal structures. *Acta Crystallogr D Biol Crystallogr* **64**, 70–5.

Nave, C. (1998): A description of imperfections in protein crystals. *Acta Crystallogr D Biol Crystallogr* **54**, 848–53.

Nave, C., and Hill, M. A. (2005): Will reduced radiation damage occur with very small crystals? *J Synch Radiat* **12**, 299–303.

Nespolo, M. (2009): Comments on the article 'Macromolecular crystal twinning, lattice disorders and multiple crystals. *Crystallogr Rev* **15**, 261–263.

Nespolo, M., and Ferraris, G. (2000): Twinning by syngonic and metric merohedry. Analysis, classification and effects on the diffraction pattern. *Z Krist* **215**, 77–81.

Neutze, R., Wouts, R., van der Spoel, D., Weckert, E., and Hajdu, J. (2000): Potential for biomolecular imaging with femtosecond X-ray pulses. *Nature* **406**, 752–7.

Newman, J. (2005): Expanding screening space through the use of alternative reservoirs in vapor-diffusion experiments. *Acta Crystallogr D Biol Crystallogr* **61**, 490–3.

Newman, J. (2006): A review of techniques for maximizing diffraction from a protein crystal in stilla. *Acta Crystallogr D Biol Crystallogr* **62**, 27–31.

Ng, J. D., Lorber, B., Giege, R., Koszelak, S., Day, J., Greenwood, A., and McPherson, A. (1997): Comparative analysis of thaumatin crystals grown on earth and in microgravity. *Acta Crystallogr D Biol Crystallogr* **53**, 724–33.

Ng, J. D., Lorber, B., Witz, J., TheobaldDietrich, A., Kern, D., and Giege, R., (1996) The crystallization of biological macromolecules from precipitates: Evidence for Ostwald ripening. *J Cryst Growth* **168, 1–4**, 50–62.

Ng, J. D., Gavira, J. A., and Garcia-Ruiz, M. A. (2003): Protein crystallization by capillary counter-diffusion for applied crystallographic structure determination. *J Struct Biol* **142**, 218–231.

Nieh, Y. P., and Helliwell, J. R. (1995): Time Differences between Friedel Reflections – Accuracy of Crystal Setting and Requirements on Beam Stability. *J Synch Radiat* **2**, 79–82.

Nieh, Y. P., Raftery, J., Weisgerber, S., Habash, J., Schotte, F., Ursby, T., Wulff, M., Hadener, A., Campbell, J. W., Hao, Q., and Helliwell, J. R. (1999): Accurate and highly complete synchrotron protein crystal Laue diffraction data using the ESRF CCD and the Daresbury Laue software. *J Synch Radiat* **6**, 995–1006.

Nield, J., Rizkallah, P. J., Barber, J., and Chayen, N. E. (2003): The 1.45 angstrom three-dimensional structure of C-phycocyanin from the thermophilic cyanobacterium Synechococcus elongatus. *J Struct Biol* **141**, 149–155.

Niimura, N., and Bau, R. (2008): Neutron protein crystallography: beyond the folding structure of biological macromolecules. *Acta Crystallogr Section A* **64**, 12–22.

Noever, D., Pusey, M. L., Forsythe, E. L., and Baskaran, S. (1996): Artificial neural network prediction of tetragonal lysozyme face growth rates. *J Cryst Growth* **167**, 221–236.

Nollert, P. (2002): From test tube to plate: a simple procedure for the rapid preparation of microcrystallization experiments using the cubic phase method. *J Appl Crystallogr* **35**, 637–640.

Norrman, M., Stahl, K., Schluckebier, G., and Al-Karadaghi, S. (2006): Characterization of insulin microcrystals using powder diffraction and multivariate data analysis. *J Appl Crystallogr* **39**, 391–400.

O'Neill, P., Stevens, D. L., and Garman, E. F. (2002): Physical and chemical considerations of damage induced in protein crystals by synchrotron radiation: a radiation chemical perspective. *J Synch Radiat* **9**, 329–332.

Ohsumi, K., Hagiya, K., and Ohmasa, M. (1991): Development of a System to Analyze the Structure of a Submicrometer-Sized Single-Crystal by Synchrotron X-Ray-Diffraction. *J Appl Crystallogr* **24**, 340–348.

Okazaki, N., Hasegawa, K., Ueno, G., Murakami, H., Kumasaka, T., and Yamamoto, M. (2008): Mail-in data collection at SPring-8 protein crystallography beamlines. *J Synch Radiat* **15**, 288–291.

Oldfield, T. J., Ceska, T. A., and Brady, R. L. (1991): A Flexible Approach to Automated Protein Crystallization. *J Appl Crystallogr* **24**, 255–260.

Onuma, K., Tsukamoto, K., and Nakadate, S. (1993): Application of Real-Time Phase-Shift Interferometer to the Measurement of Concentration Field. *J Cryst Growth* **129**, 706–718.

Ostermeier, C., Iwata, S., Ludwig, B., and Michel, H. (1995): Fv fragment-mediated crystallization of the membrane protein bacterial cytochrome c oxidase. *Nature Struct Biol* **2**, 842–6.

Ostwald, W. (1896): *Lehrbuch der Allgemeinen Chemie*. Leipzig, Germany.

Otalora, F., Garcia Ruiz, J. M., Garvira, J. A., and Capelle, B. (1999): Topography and high resolution diffraction studies in tetragonal lysozyme. *J Cryst Growth* **196**, 546–558.

Otalora, F., Novella, M. L., Gavira, J. A., Thomas, B. R., and Garcia Ruiz, J. M. (2001): Experimental evidence for the stability of the depletion zone around a growing protein crystal under microgravity. *Acta Crystallogr D Biol Crystallogr* **57**, 412–7.

Otwinowski, Z., and Minor, W. (1997): Processing of X-ray Diffraction Data Collected in Oscillation Mode, pp. 307–326. In C. W. Carter Jr., and R. M. Sweet (ed.): *Methods in Enzymology*, Academic Press, New York.

Owen, R. L., Rudino-Pinera, E., and Garman, E. F. (2006): Experimental determination of the radiation dose limit for cryocooled protein crystals. *Proc Nat Acad Sci USA* **103**, 4912–4917.

Padilla, J. E., and Yeates, T. O. (2003): A statistic for local intensity differences: robustness to anisotropy and pseudo-centering and utility for detecting twinning. *Acta Crystallogr D Biol Crystallogr* **59**, 1124–1130.

Page, R., and Stevens, R. C. (2004): Crystallization data mining in structural genomics: using positive and negative results to optimize protein crystallization screens. *Methods* **34**, 373–389.

Pagola, S., and Stephens, P. W. (2000): Towards the solution of organic crystal structures by powder diffraction. *European Powder Diffraction, Pts 1 and 2* **321–3**, 40–45.

Parkin, S., and Hope, H. (1998): Macromolecular cryocrystallography: Cooling, mounting, storage and transportation of crystals. *J Appl Crystallogr* **31**, 945–953.

Parsons, S. (2003): Introduction to twinning. *Acta Crystallogr D Biol Crystallogr* **59**, 1995–2003.

Pechkova, E., and Nicolini, C. (2004a): Atomic structure of a CK2alpha human kinase by microfocus diffraction of extra-small microcrystals grown with nanobiofilm template. *J Cell Biochem* **91**, 1010–20.

Pechkova, E., and Nicolini, C. (2004b): Protein nanocrystallography: a new approach to structural proteomics. *Trends Biotechnol* **22**, 117–22.

Perrakis, A., Cipriani, F., Castagna, J. C., Claustre, L., Burghammer, M., Riekel, C., and Cusack, S. (1999): Protein microcrystals and the design of a microdiffractometer: current experience and plans at EMBL and ESRF/ID13. *Acta Crystallogr D Biol Crystallogr* **55**, 1765–1770.

Perutz, M. F. (1946): The Composition and Swelling Properties of Haemoglobin Crystals. *Trans. Faraday. Soc.* **42B**, 187–195.

Peyridieu, J. F., Baudot, A., Boutron, P., Mazuer, J., Odin, J., Ray, A., Chapelier, E., Payen, E., and Descotes, J. L. (1996): Critical cooling and warming rates to avoid ice crystallization in small pieces of mammalian organs permeated with cryoprotective agents. *Cryobiology* **33**, 436–446.

Pflugrath, J. W. (1999): The finer things in X-ray diffraction data collection. *Acta Crystallogr D Biol Crystallogr* **55**, 1718–1725.

Pflugrath, J. W. (2004): Macromolecular cryocrystallography – methods for cooling and mounting protein crystals at cryogenic temperatures. *Methods* **34**, 415–423.

Pickersgill, R. W. (1987): One-Dimensional Disorder in Spinach Ribulose Bisphosphate Carboxylase Crystals. *Acta Crystallogr Section A* **43**, 502–506.

Pickford, M. G., Garman, E. F., Jones, E. Y., and Stuart, D. I. (1993): A Design of Crystal Mounting Cell That Allows the Controlled Variation of Humidity at the Protein Crystal during X-Ray-Diffraction. *J Appl Crystallogr* **26**, 465–466.

Plohn, H. J., and Buldt, G. (1986): The Determination of Label Positions in Membrane-Proteins by Neutron and Anomalous X-Ray-Diffraction of Powder Samples. *J Appl Crystallogr* **19**, 255–261.

Plomp, M., McPherson, A., and Malkin, A. J. (2003): Repair of impurity-poisoned protein crystal surfaces. *Proteins* **50**, 486–95.

Polykarpov, I., and Sawyer, L. (1995): Correction on perfection: primary extinction correction in protein crystallography. *Joint CCP4 + ESF-EAMCB Newsletter on Protein Crystallography* **31**, 5–11; STFC, Daresbury, UK.

Poodt, P. W. G., Heijna, M. C. R., Christianen, P. C. M., van Enckevort, W. J. P., de Grip, W. J., Tsukamoto, K., Maan, J. C., and Vlieg, E. (2006): Using gradient magnetic fields to suppress convection during crystal growth. *Cryst Growth & Design* **6**, 2275–2280.

Priebe, G., Laundy, D., MacDonald, M. A., Diakun, G. P., Jamison, S. P., Jones, L. B., Holder, D. J., Smith, S. L., Phillips, P. J., Fell, B. D., Sheehy, B., Naumova, N., Sokolov, I. V., Ter-Avetisyan, S., Spohr, K., Krafft, G. A., Rosenzweig, J. B., Schramm, U., Gruner, F., Hirst, G. J., Collier, J., Chattopadhyay, S., and Seddon, E. A. (2008): Inverse Compton backscattering source driven by the multi-10 TW laser installed at Daresbury. *Laser and Particle Beams* **26**, 649–660.

Pusey, M. L., Witherow, W., and Naumann, R. (1988): Preliminary investigations into solutal flow about growing tetragonal lysozyme crystals. *J. Cryst Growth* **90**, 105–111.

Putnam, C. D., Hammel, M., Hura, G. L., and Tainer, J. A. (2007): X-ray solution scattering (SAXS) combined with crystallography and computation: defining accurate macromolecular structures, conformations and assemblies in solution. *Q Rev Biophys* **40**, 191–285.

Qui, X., and Janson, C. A. (2009): Improving Protein Crystallizability by Modifications and Engineering, pp. 249–266. In T. M. Bergfors (ed.): *Protein crystallization*, International University Line, La Jolla, California.

Ramakrishnan, V., and Biou, V. (1997): Treatment of multiwavelength anomalous diffraction data as a special case of multiple isomorphous replacement. *Macromolec Crystallogr, Pt A* **276**, 538–557.

Ravelli, R. B. G., and McSweeney, S. M. (2000): The 'fingerprint' that X-rays can leave on structures. *Structure* **8**, 315–328.

Rayment, I. (1997): Reductive alkylation of lysine residues to alter crystallization properties of proteins. *Macromolec Crystallogr, Pt A* **276**, 171–179.

Reckenthaeler, P., Centurion, M., Fuß, W., Trushin, S. A., Krausz, F., and Fill, E. E., (2009). Time resolved electron diffraction from selectively aligned molecules. *Phys Rev Lett* **102**, 213001.

Rees, D. C. (1980): The Influence of Twinning by Merohedry on Intensity Statistics. *Acta Crystallogr Section A* **36**, 578–581.

Rees, D. C. (1982): A General-Theory of X-Ray-Intensity Statistics for Twins by Merohedry. *Acta Crystallogr Section A* **38**, 201–207.

Ren, J., Esnouf, R., Garman, E., Somers, D., Ross, C., Kirby, I., Keeling, J., Darby, G., Jones, Y., Stuart, D., and *et al.* (1995): High resolution structures of HIV-1 RT from four RT-inhibitor complexes. *Nature Struct Biol* **2**, 293–302.

Ren, Z., Bourgeois, D., Helliwell, J. R., Moffat, K., Srajer, V., and Stoddard, B. L. (1999): Laue crystallography: coming of age. *J Synch Radiat* **6**, 891–917.

Ren, Z., Perman, B., Srajer, V., Teng, T. Y., Pradervand, C., Bourgeois, D., Schotte, F., Ursby, T., Kort, R., Wulff, M., and Moffat, K. (2001): A molecular movie at 1.8 angstrom resolution displays the photocycle of photoactive yellow protein, a eubacterial blue-light receptor, from nanoseconds to seconds. *Biochemistry* **40**, 13788–13801.

Renault, L., Hanzal-Bayer, M., and Hillig, R. C. (2001): Co-expression, copurification, crystallization and preliminary X-ray analysis of a complex of ARL2-GTP and PDE delta. *Acta Crystallogr D Biol Crystallogr* **57**, 1167–1170.

Rieck, W., Euler, H., Schulz, H., and Schildkamp, W. (1988): Synchrotron X-Ray-Diffraction on a CaF_2 Microcrystal with 2.2 Cubic Micrometers Volume. *Acta Crystallogr Section A* **44**, 1099–1101.

Ries-Kautt, M. (1999), pp. 91–109. In T. M. Bergfors (ed.): *Protein crystallization techniques,strategies and tips*, International University Line, California.

Robert, M. C., and Lefaucheux, F. (1988): Crystal growth in gels: Principle and applications. *J. Cryst Growth* **90**, 358–367.

Roberts, B. R., Wood, Z. A., Jonsson, T. J., Poole, L. B., and Karplus, P. A. (2005): Oxidized and synchrotron cleaved structures of the disulfide redox center in the N-terminal domain of Salmonella typhimurium AhpF. *Protein Sci* **14**, 2414–2420.

Rodgers, D. W. (1994): Cryocrystallography. *Structure* **2**, 1135–40.

Rodgers, D. W. (1997): Practical Cryocrystallography, pp. 183–203. In C. W. J. Carter, and R. M. Sweet (ed.): *Macromolec Crystallogr Part B*, Academic Press, New York.

Rollet, J. S. (1965): *Computing methods in crystallography*. Pergamon. Oxford.

Rosenberger, F., Howard, S. B., Sowers, J. W., and Nyce, T. A. (1993): Temperature-dependance of protein solubility – Determination and application to crystallization in X-ray capillaries. *J. Cryst Growth* **129**, 1–12.

Rossmann, M. G., and Arnold, E. (2001): Crystallography of Biological Macromolecules: *International tables for crystallography*, Kluwer Academic Publishers, Dordrecht.

Rubin, B., Talafous, J., and Larson, D. (1991): Minimal Intervention Robotic Protein Crystallization. *J Cryst Growth* **110**, 156–163.

Rudolph, M. G., Kelker, M. S., Schneider, T. R., Yeates, T. O., Oseroff, V., Heidary, D. K., Jennings, P. A., and Wilson, I. A. (2003): Use of multiple anomalous dispersion to phase highly merohedrally twinned crystals of interleukin-1 beta. *Acta Crystallogr D Biol Crystallogr* **59**, 290–298.

Rupp, B., and Wang, J. W. (2004): Predictive models for protein crystallization. *Methods* **34**, 390–407.

Rypniewski, W. R., Holden, H. M., and Rayment, I. (1993): Structural consequences of reductive methylation of lysine residues in hen egg white lysozyme: an X-ray analysis at 1.8-A resolution. *Biochemistry* **32**, 9851–8.

Sadaoui, N., Janin, J., and Lewitbentley, A. (1994): Taos – an Automatic System for Protein Crystallization. *J Appl Crystallogr* **27**, 622–626.

Saijo, S., Yamada, Y., Sato, T., Tanaka, N., Matsui, T., Sazaki, G., Nakajima, K., and Matsuura, Y. (2005b): Structural consequences of hen egg-white lysozyme orthorhombic crystal growth in a high magnetic field: validation of X-ray diffraction intensity, conformational energy searching and quantitative analysis of B factors and mosaicity. *Acta Crystallogr D Biol Crystallogr* **61**, 207–17.

Salunke, D. M. (2007): Comment on experimental parameters for humidity experiments.

Salunke, D. M., Veerapandian, B., Kodandapani, R., and Vijayan, M. (1985): Water-Mediated Transformations in Protein Crystals. *Acta Crystallogr Section B-Structural Science* **41**, 431–436.

Salunke, D. M., Veerpandian, B., and Vijayan, M. (1984): Water-Mediated Structural Transformations in a New Crystal Form of Ribonuclease-a and Tetragonal Lysozyme. *Current Science* **53**, 231–235.

Samygina, V. R., Antonyuk, S. V., Lamzin, V. S., and Popov, A. N. (2000): Improving the X-ray resolution by reversible flash-cooling combined with concentration screening, as exemplified with PPase. *Acta Crystallogr D Biol Crystallogr* **56**, 595–603.

Saric, M., Zhao, X., Korner, C., Nowak, C., Kuhlmann, J., and Vetter, I. R. (2007): Structural and biochemical characterization of the Importin beta.Ran.GTP. RanBD1 complex, *FEBS Lett.* **581**, 1369–1376.

Saridakis, E., and Chayen, N. E. (2003): Systematic improvement of protein crystals by determining the supersolubility curves of phase diagrams. *Biophys J* **84**, 1218–1222.

Saridakis, E., and Chayen, N. E. (2009): Towards a 'Universal' Nucleant for Protein Crystallization, *Trends Biotechnol* **27**, 99–106.

Saridakis, E., Dierks, K., Moreno, A., Dieckmann, M. W. M., and Chayen, N. E. (2002): Separating nucleation and growth in protein crystallization using dynamic light scattering. *Acta Crystallogr D Biol Crystallogr* **58**, 1597–1600.

Saridakis, E. E. G., Stewart, P. D. S., Lloyd, L. F., and Blow, D. M. (1994): Phase-Diagram and Dilution Experiments in the Crystallization of Carboxypeptidase-G (2). *Acta Crystallogr D Biol Crystallogr* **50**, 293–297.

Sazaki, G., Yoshida, E., Komatsu, H., Nakada, T., Miyashita, S., and Watanabe, K. (1997): Effects of a magnetic field on the nucleation and growth of protein crystals. *J Cryst Growth* **173**, 231–234.

Scheuler, J., Frank, J., Saenger, W., and Georgalis, Y. (1999): Thermally induced aggregation of human transferring receptor studied by light-scattering techniques. *Biophys J* **77**, 1117–1125.

Schiltz, M., Kvick, A., Svensson, O. S., Shepard, W., delaFortelle, E., Prange, T., Kahn, R., Bricogne, G., and Fourme, R. (1997): Protein crystallography at ultra-short wavelengths: Feasibility study of anomalous-dispersion experiments at the xenon K-edge. *J Synch Radiat* **4**, 287–297.

Schmidt, K., Spence, J., Weierstall, U., Starodub, D., Chapman, H., Howells, M., and Doak, B. (2008): Triple-beam tomographic femtosecond X-ray diffractive imaging. *Phys Rev Lett* **101**, 115507.

Schmitz, S. K. (1990): *An introduction to dynamic light scattering by macromolecules*. Academic Press. New York.

Schneider, T. R., Karcher, J., Pohl, E., Lubini, P., and Sheldrick, G. M. (2000): Ab initio structure determination of the lantibiotic mersacidin. *Acta Crystallogr D Biol Crystallogr* **56**, 705–13.

Schubot, F. D., and Waugh, D. S. (2004): A pivotal role for reductive methylation in the de novo crystallization of a ternary complex composed of Yersinia pestis virulence factors YopN, SycN and YscB. *Acta Crystallogr D Biol Crystallogr* **60**, 1981–6.

Segelke, B. W. (2001): Efficiency analysis of sampling protocols used in protein crystallization screening. *J Cryst Growth* **232**, 553–562.

Shaikevitch, A., and Kam, Z. (1981): Investigation of long-range order in protein crystals by X-ray diffraction. *Acta Crystallogr A* **37**, 871–875.

Shankland, K., David, W. I. F., and Csoka, T. (1997a): Crystal structure determination from powder diffraction data by the application of a genetic algorithm. *Zeitschrift Fur Kristallographie* **212**, 550–552.

Shankland, K., David, W. I. F., and Sivia, D. S. (1997b): Routine ab initio structure determination of chlorothiazide by X-ray powder diffraction using optimised data collection and analysis strategies. *J Materials Chemistry* **7**, 569–572.

Shapiro, D. A., De Ponte, D., Doak, R. B., Fromme, P., Hembree, G., Hunter, M., Marchesini, S., Schmidt, K., Weierstall, U., and Spence, J. C. H. (2008): Powder diffraction from a continuous microjet of nanoscale protein crystals. *J Synch Radiat* **15**, 593–599.

Shaw, N., Cheng, C., Tempel, W., Chang, J., Ng, J., Wang, X. Y., Perrett, S., Rose, J., Rao, Z., Wang, B. C., and Liu, Z. J. (2007): (NZ)CH...O contacts assist crystallization of a ParB-like nuclease. *BMC Struct Biol* **7**, 46.

Shieh, H. S., Stallings, W. C., Stevens, A. M., and Stegeman, R. A. (1995): Using Sampling Techniques in Protein Crystallization. *Acta Crystallogr D Biol Crystallogr* **51**, 305–310.

Shimizu, N., Hirata, K., Hasegawa, K., Ueno, G., and Yamamoto, M. (2007): Dose dependence of radiation damage for protein crystals studied at various X-ray energies. *J Synch Radiat* **14**, 4–10.

Shlichta, P. J. (1986): Feasibility of Mapping Solution Properties during the Growth of Protein Crystals. *J Cryst Growth* **76**, 656–662.

Shu, F., Ramakrishnan, V., and Schoenborn, B. P. (2000): Enhanced visibility of hydrogen atoms by neutron crystallography on fully deuterated myoglobin. *Proc Natl Acad Sci U S A* **97**, 3872–7.

Sica, F., Demasi, D., Mazzarella, L., Zagari, A., Capasso, S., Pearl, L. H., Dauria, S., Raia, C. A., and Rossi, M. (1994): Elimination of Twinning in Crystals of Sulfolobus-Solfataricus Alcohol-Dehydrogenase Holo-Enzyme by Growth in Agarose Gels. *Acta Crystallogr D Biol Crystallogr* **50**, 508–511.

Simonovic, M., and Volz, K. (2002): Atomic resolution structure of a succinimide intermediate in E.coli CheY. *J Mol Biol* **322**, 663–7.

Singer, W., Rubinsztein-Dunlop, H., and Gibson, U. (2004): Manipulation and growth of birefringent protein crystals in optical tweezers. *Optics Express* **12**, 6440–6445.

Sjogren, T., Carlsson, G., Larsson, G., Hajdu, A., Andersson, C., Pettersson, H., and Hajdu, J. (2002): Protein crystallography in a vapour stream: data collection, reaction initiation and intermediate trapping in naked hydrated protein crystals. *J Appl Crystallogr* **35**, 113–116.

Skinner, J. M., Cowan, M., Buono, R., Nolan, W., Bosshard, H., Robinson, H. H., Heroux, A., Soares, A. S., Schneider, D. K., and Sweet, R. M. (2006): Integrated software for macromolecular crystallography synchrotron beamlines II: revision, robots and a database. *Acta Crystallogr D Biol Crystallogr* **62**, 1340–1347.

Skouri, M., Munch, J. P., Lorber, B., Giege, R., and Candau, S. (1992): Interactions between Lysozyme Molecules under Precrystallization Conditions Studied by Light-Scattering. *J Cryst Growth* **122**, 14–20.

Slabinski, L., Jaroszewski, L., Rodrigues, A. P. C., Rychlewski, L., Wilson, I. A., Lesley, S. A., and Godzik, A. (2007a): The challenge of protein structure determination – lessons from structural genomics. *Protein Sci* **16**, 2472–2482.

Slabinski, L., Jaroszewski, L., Rychlewski, L., Wilson, I. A., Lesley, S. A., and Godzik, A. (2007b): XtalPred: a web server for prediction of protein crystallizability. *Bioinformatics* **23**, 3403–3405.

Smatanova, I. K., Gavira, J. A., Rezacova, P., Vacha, F., and Garcia-Ruiz, J. M. (2006): New techniques for membrane protein crystallization tested on photosystem II core complex of Pisum sativum. *Photosynthesis Research* **90**, 255–259.

Snell, E. H. (1995): Uses of the synchrotron X-ray Laue technique and investigation of microgravity crystal growth, pp. 296: *Chemistry*, Ph.D. Thesis, University of Manchester.

Snell, E. H., Bellamy, H. D., Rosenbaum, G., and van der Woerd, M. J. (2007): Non-invasive measurement of X-ray beam heating on a surrogate crystal sample. *J Synch Radiat* **14**, 109–115.

Snell, E. H., Borgstahl, G. E., and Bellamy, H. (2003): Macromolecular Crystal Quality, pp. 268–288. In C. W. Carter Jr., and R. M. Sweet (ed.): *Macromolec Crystallogr part C*.

Snell, E. H., Cassetta, A., Helliwell, J. R., Boggon, T. J., Chayen, N. E., Weckert, E., Hölzer, K., Schroer, K., Gordon, E. J., and Zagalsky, P. F. (1997): Partial Improvement of Crystal Quality for Microgravity-Grown Apocrustacyanin C_1. *Acta Crystallogr D Biol Crystallogr* **53**, 231–239.

Snell, E. H., and Helliwell, J. R. (2005): Macromolecular crystallization in microgravity. *Rep. Prog. Phys.* **68**, 799–853.

Snell, E. H., Helliwell, J. R., Boggon, T. J., Lautenschlager, P., and Potthast, L. (1996): Lysozyme crystal growth kinetics monitored using a Mach–Zehnder interferometer. *Acta Crystallogr D Biol Crystallogr* **52**, 529–533.

Snell, E. H., Judge, R.A., Crawford, L., Forsythe, E.L., Pusey, M.L., Sportiello, M., Todd, P., Bellamy, H., Lovelace, J., Cassanto, Borgstahl, G.E.O (2001): Impurity partitioning electrophoretic and mosaicity analysis of microgravity and ground-grown crystals. *Cryst Growth and Design* **1**, 151–158.

Snell, E. H., Nagel, R. M., Wojtaszcyk, A., Wolfley, J., O'Neill, H., and Luft, J. R. (2008): The application and use of Chemical Space Mapping to interpret crystallization screening results. *Acta Crystallogr D.* **64**, 1240–1249.

Snell, E. H., van der Woerd, M. J., Damon, M., Judge, R. A., Myles, D. A., and Meilleur, F. (2006): Optimizing crystal volume for neutron diffraction: D-xylose isomerase. *Eur Biophys J* **35**, 621–32.

Snell, E. H., Weisgerber, S., Helliwell, J. R., Holzer, K., and Schroer, K. (1995): Improvements in lysozyme protein crystal perfection through microgravity growth. *Acta Crystallogr D Biol Crystallogr* **51**, 1099–1102.

Snijder, A., Barrends, T. & Dijkstra, D (2003): Crystallization of Phospholipase A in Two Biological Oligomerization states., pp. 265–278. In S. Iwata (ed.): *Methods and results in crystallization of membrane proteins* International University Line, La Jolla, California.

Snijder, H. J., Timmins, P. A., Kalk, K. H., and Dijkstra, B. W. (2003): Detergent organisation in crystals of monomeric outer membrane phospholipase A. *J Struct Biol* **141**, 122–131.

Snyder, D. A., Chen, Y., Denissova, N. G., Acton, T., Aramini, J. M., Ciano, M., Karlin, R., Liu, J., Manor, P., Rajan, P. A., Rossi, P., Swapna, G. V., Xiao, R., Rost, B., Hunt, J., and Montelione, G. T. (2005): Comparisons of NMR spectral quality and success in crystallization demonstrate that NMR and X-ray crystallography are complementary methods for small protein structure determination. *J Am Chem Soc* **127**, 16505–11.

Soltis, S. M., Cohen, A. E., Deacon, A., Eriksson, T., Gonzalez, A., McPhillips, S., Chui, H., Dunten, P., Hollenbeck, M., Mathews, I., Miller, M., Moorhead, P., Phizackerley, R. P., Smith, C., Song, J., van dem Bedem, H., Ellis, P., Kuhn, P., McPhillips, T., Sauter, N., Sharp, K., Tsyba, I., and Wolf, G. (2008): New paradigm for macromolecular crystallography experiments at SSRL: automated crystal screening and remote data collection. *Acta Crystallogr D Biol Crystallogr* **64**, 1210–1221.

Sommer, M. O. A., and Larsen, S. (2005): Crystallizing proteins on the basis of their precipitation diagram determined using a microfluidic formulator. *J Synch Radiat* **12**, 779–785.

Song, H., and Ismagilov, R. F. (2003): Millisecond kinetics on a microfluidic chip using nanoliters of reagents. *J Am Chem Soc* **125**, 14613–14619.

Soriano, T. M. B., and Fontecillacamps, J. C. (1993): Astec – an Automated-System for Sitting-Drop Protein Crystallization. *J Appl Crystallogr* **26**, 558–562.

Southworth-Davies, R. J., Medina, M. A., Carmichael, I., and Garman, E. F. (2007): Observation of decreased radiation damage at higher dose rates in room temperature protein crystallography. *Structure* **15**, 1531–41.

Spence, J. C. H. (2007): Lensless (diffractive) Imaging. Chapter in *Science of microscopy*. ed. P. Hawkes and J. C. H. Spence. (Springer. 2 Vols. New York).

Spence, J. C., Schmidt, K., Wu, J. S., Hembree, G., Weierstall, U., Doak, B., and Fromme, P. (2005): Diffraction and imaging from a beam of laser-aligned proteins: resolution limits. *Acta Crystallogr A* **61**, 237–45.

Srajer, V., Ren, Z., Teng, T. Y., Schmidt, M., Ursby, T., Bourgeois, D., Pradervand, C., Schildkamp, W., Wulff, M., and Moffat, K. (2001): Protein conformational relaxation and ligand migration in myoglobin: A nanosecond to millisecond molecular movie from time-resolved Laue X-ray diffraction. *Biochemistry* **40**, 13802–13815.

Srajer, V., Teng, T. Y., Ursby, T., Pradervand, C., Ren, Z., Adachi, S., Schildkamp, W., Bourgeois, D., Wulff, M., and Moffat, K. (1996): Photolysis of the carbon monoxide complex of myoglobin: Nanosecond time-resolved crystallography. *Science* **274**, 1726–1729.

Stanley, E. (1972): Identification of Twins from Intensity Statistics. *J Appl Crystallogr* **5**, 191–194.

Stocker, A., Keis, S., Cook, G. M., and Dimroth, P. (2005): Purification, crystallization, and properties of F1-ATPase complexes from the thermoalkaliphilic Bacillus sp. strain TA2.A1. *J Struct Biol* **152**, 140–145.

Stojanoff, V., and Siddons, D. P. (1996): X-ray topography of a lysozyme crystal. *Acta Crystallogr A* **52**, 498–499.

Stojanoff, V., Siddons, D. P., Monaco, L. A., Vekilov, P., and Rosenberger, F. (1997): X-ray Topography of Tetragonal Lysozyme Grown by the Temperature-Controlled Technique. *Acta Crystallogr D Biol Crystallogr* **53**, 588–595.

Stojanoff, V., Siddons, D. P., Snell, E. H., and Helliwell, J. R. (1996): X-ray topography: An old technique with a new application. *Synch Radiat News* **9**, 25–26.

Stura, E. A. (1999): Seeding, pp. 141–153. In T. Bergfors (ed.): *Protein crystallization: techniques, strategies and tips.*, International University Line, La Jolla.

Stura, E. A., and Wilson, I. A. (1991): Applications of the Streak Seeding Technique in Protein Crystallization. *J Cryst Growth* **110**, 270–282.

Swatkoski, S., Gutierrez, P., Ginter, J., Petrov, A., Dinman, J. D., Edwards, N., and Fenselau, C. (2007): Integration of residue-specific acid cleavage into proteomic workflows. *J Proteome Res* **6**, 4525–7.

Takeda, K., Miyatake, H., Park, Y., Kawamoto, M., Kamiya, N., and Miki, K (2004): Multi-wavelength anomalous diffractin mehtod for I and Xe atoms using ultra-high-energy X-rays from SPring-8. *J Appl Crystallogr* **37**, 925.

Tanaka, H., Inaka, K., Sugiyama, S., Takahashi, S., Sano, S., Sato, M., and Yoshitomi, S. (2004): A simplified counter diffusion method combined with a 1-D simulation program for optimizing crystallization conditions. *J Synch Radiat* **11**, 45–48.

Tanaka, S., and Ataka, M. (2002): Protein crystallization induced by polyethylene glycol: A model study using apoferritin. *J Chemical Physics* **117**, 3504–3510.

Taylor, G. (2003): The phase problem. *Acta Crystallogr D Biol Crystallogr* **59**, 1881–1890.

ten Wolde, P. R., and Frenkel, D. (1997): Enhancement of protein crystal nucleation by critical density fluctuations. *Science* **277**, 1975–8.

Teng, T. Y. (1990): Mounting crystals for macromolecular crystallography in a free-standing thin film. *J Appl Crystallogr* **23**, 387–391.

Teng, T. Y., and Moffat, K. (1998): Cooling rates during flash cooling. *J Appl Crystallogr* **31**, 252–257.

Tereshko, V., Uysal, S., Koide, A., Margalef, K., Koide, S., and Kossiakoff, A. A. (2008): Toward chaperone-assisted crystallography: protein engineering enhancement of crystal packing and X-ray phasing capabilities of a camelid single-domain antibody (VHH) scaffold. *Protein Sci* **17**, 1175–87.

Terwisscha van Scheltinga, A. C., Valegard, K., Hajdu, J., and Andersson, I. (2003): MIR phasing using merohedrally twinned crystals. *Acta Crystallogr D Biol Crystallogr* **59**, 2017–22.

Thorne, R. E., Stum, Z., Kmetko, J., O'Neill, K., and Gillilan, R. (2003): Microfabricated mounts for high-throughput macromolecular crystallography. *J Appl Crystallogr* **36**, 1455–1460.

Thorsen, T., Roberts, R. W., Arnold, F. H., and Quake, S. R. (2001): Dynamic pattern formation in a vesicle-generating microfluidic device. *Physical Review Letters* **86**, 4163–4166.

Timmins, P. (2006): Detergent Binding in Membrane Protein Crystals by Neutron Crystallography, pp. 73–82. In J. Fitter, T. Gutberlet, and J. Katsaras (ed.): *Neutron scattering in biology*, Springer-Verlag, Berlin, Heidelberg, Germany.

Tsutakawa, S. E., Hura, G. L., Frankel, K. A., Cooper, P. K., and Tainer, J. A. (2007): Structural analysis of flexible proteins in solution by small angle X-ray scattering combined with crystallography. *J Struct Biol* **158**, 214–23.

Tzakos, A. G., Grace, C. R., Lukavsky, P. J., and Riek, R. (2006): NMR techniques for very large proteins and rnas in solution. *Annu Rev Biophys Biomol Struct* **35**, 319–42.

Uson, I., and Sheldrick, G. M. (1999): Advances in direct methods for protein crystallography. *Curr Opin Struct Biol* **9**, 643–648.

Vahedi-Faridi, A., Lovelace, J., Bellamy, H. D., Snell, E. H., and Borgstahl, G. E. (2003): Physical and structural studies on the cryocooling of insulin crystals. *Acta Crystallogr D Biol Crystallogr* **59**, 2169–82.

van Smaalen, S. (2007): *Incommensurate Crystallography*. Oxford University Press, Oxford, UK.

van Thor, J. J., Gensch, T., Hellingwerf, K. J., and Johnson, L. N. (2002): Phototransformation of green fluorescent protein with UV and visible light leads to decarboxylation of glutamate 222. *Nature Struct Biol* **9**, 37–41.

Veesler, S., Marcq, S., Lafont, S., Astier, J. P., and Boistelle, R. (1994): Influence of Polydispersity on Protein Crystallization – a Quasi-Elastic Light-Scattering Study Applied to Alpha-Amylase. *Acta Crystallogr D Biol Crystallogr* **50**, 355–360.

Vekilov, P. G., Ataka, M., and Katsura, T. (1995): Growth-Processes of Protein Crystals Revealed by Laser Michelson Interferometry Investigation. *Acta Crystallogr D Biol Crystallogr* **51**, 207–219.

Volz, H. M., and Matyi, R. J. (1999): High-resolution X-ray diffraction analyses of protein crystals. *Philos Trans. R. Soc. Lond. A* **357**, 2789–2799.

Volz, H. M., and Matyi, R. J. (2000): Triple-axis X-ray diffraction analyses of lysozyme crystals. *Acta Crystallogr D Biol Crystallogr* **56**, 881–9.

Volz, H. M., and Matyi, R. J. (2001): A high resolution triple axis X-ray diffraction analysis of radiation damage in lysozyme crystals. *J Cryst Growth* **232**, 502–510.

Von Dreele, R. B. (2007): Multipattern Rietveld refinement of protein powder data: an approach to higher resolution. *J Appl Crystallogr* **40**, 133–143.

Von Dreele, R. B., Lee, P. L., and Zhang, Y. (2006): Protein polycrystallography. *Zeitschrift Fur Kristallographie* **Sup**, 3–8.

Von Dreele, R. B., Stephens, P. W., Smith, G. D., and Blessing, R. H. (2000): The first protein crystal structure determined from high-resolution X-ray powder diffraction data: a variant of T3R3 human insulin-zinc complex produced by grinding. *Acta Crystallogr D Biol Crystallogr* **56**, 1549–1553.

Wakayama, N. I., Yin, D. C., Harata, K., Kiyoshi, T., Fujiwara, M., and Tanimoto, Y. (2006): Macromolecular crystallization in microgravity generated by a superconducting magnet. *Ann N Y Acad Sci* **1077**, 184–93.

Walker, C. G., Foadi, J., and Wilson, J. (2007): Classification of protein crystallization images using Fourier descriptors. *J Appl Crystallogr* **40**, 418–426.

Walker, L. J., Moreno, P. O., and Hope, H. (1998): Cryocrystallography: effect of cooling medium on sample cooling rate. *J Appl Crystallogr* **31**, 954–956.

Wall, M. E., Ealick, S. E., and Gruner, S. M. (1997): Three-dimensional diffuse X-ray scattering from crystals of Staphylococcal nuclease. *Proc Nat Acad Sci USA* **94**, 6180–6184.

Walter, T. S., Diprose, J., Brown, J., Pickford, M., Owens, R. J., Stuart, D. I., and Harlos, K. (2003): A procedure for setting up high-throughput nanolitre crystallization experiments. I. Protocol design and validation. *J Appl Crystallogr* **36**, 308–314.

Walter, T. S., Mancini, E. J., Kadlec, J., Graham, S. C., Assenberg, R., Ren, J., Sainsbury, S., Owens, R. J., Stuart, D. I., Grimes, J. M., and Harlos, K. (2008): Semi-automated microseeding of nanolitre crystallization experiments. *Acta Crystallogr Section F-Structural Biology and Crystallization Communications* **64**, 14–18.

Walter, T. S., Meier, C., Assenberg, R., Au, K. F., Ren, J., Verma, A., Nettleship, J. E., Owens, R. J., Stuart, D. I., and Grimes, J. M. (2006): Lysine methylation as a routine rescue strategy for protein crystallization. *Structure* **14**, 1617–22.

Wang, B. C. (1985): Resolution of Phase Ambiguity in Macromolecular Crystallography, pp. 90–112. In H. W. Wyckoff, C. H. W. Hirs, and S. N. Timasheff (ed.): *Diffraction methods for biological macromolecules*, Academic Press, San Diego, USA.

Wang, J., Rho, S. H., Park, H. H., and Eom, S. H. (2005): Correction of X-ray intensities from an HslV-HslU co-crystal containing lattice-translocation defects. *Acta Crystallogr D Biol Crystallogr* **61**, 932–41.

Ward, K. B., Perozzo, M. A., and Zuk, W. M. (1988): Automatic Preparation of Protein Crystals Using Laboratory Robotics and Automated Visual Inspection. *J Cryst Growth* **90**, 325–339.

Warren, J. E., Diakun, G., Bushnell-Wye, G., Fisher, S., Thalal, A., Helliwell, M., and Helliwell, J. R. (2008): Science experiments via telepresence at a synchrotron radiation source facility. *J Synch Radiat* **15**, 191–194.

Weckert, E., and Hummer, K. (1997): Multiple-beam X-ray diffraction for physical determination of reflection phases and its applications. *Acta Crystallogr A* **53**, 108–143.

Weik, M., Berges, J., Raves, M. L., Gros, P., McSweeney, S., Silman, I., Sussman, J. L., Houee-Levin, C., and Ravelli, R. B. G. (2002): Evidence for the formation of disulfide radicals in protein crystals upon X-ray irradiation. *J Synch Radiat* **9**, 342–346.

Weik, M., Ravelli, R. B. G., Kryger, G., McSweeney, S., Raves, M. L., Harel, M., Gros, P., Silman, I., Kroon, J., and Sussman, J. L. (2000): Specific chemical and structural damage to proteins produced by synchrotron radiation. *Proc Nat Acad Sci USA* **97**, 623–628.

Weisgerber, S., and Helliwell, J. R. (1993a): Chicken egg-white lysozyme crystals grown in microgravity and on earth for comparison of their respective perfection. *Joint CCP4 + ESF-EAMCB Newsletter on Protein Crystallography* **29**, 10–13; STFC, Daresbury, UK.

Weisgerber, S., and Helliwell, J. R. (1993b): High-resolution crystallographic studies of native concanavalin A using rapid Laue data methods and the introduction of a monochromatic large-angle oscillation technique (LOT). *J. Chem. Soc. Faraday Trans.* **89**, 2667–2675.

Weiss, M. S. (2001): Global indicators of X-ray data quality. *J Appl Crystallogr* **34**, 130–135.

Weiss, M. S., and Hilgenfeld, R. (1999): Dehydration leads to a phase transition in monoclinic factor XIII crystals. *Acta Crystallogr D Biol Crystallogr* **55**, 1858–1862.

Weiss, M. S., Panjikar, S., Mueller-Dieckmann, C., and Tucker, P. A. (2005): On the influence of the incident photon energy on the radiation damage in crystalline biological samples. *J Synch Radiat* **12**, 304–309.

Welberry, T. R. (2004): *Diffuse X-ray scattering and models of disorder*. Oxford University Press, Oxford, UK.

Wernimont, A., and Edwards, A. (2009): In situ proteolysis to generate crystals for structure determination: an update. *PLoS ONE* **4**, e5094.

White, S. H. (2004): The progress of membrane protein structure determination. *Protein Sci* **13**, 1948–1949.

Wiener, M. C. (2004): A pedestrian guide to membrane protein crystallization. *Methods* **34**, 364–372.

Wilson, A. J. C. (1942): Determination of absolute from relative X-ray intensity data. *Nature* **150**, 151–152.

Wilson, J. (2002): Towards the automated evaluation of crystallization trials. *Acta Crystallogr D Biol Crystallogr* **58**, 1907–1914.

Wilson, J., and Main, P. (2000): The use of wavelet transforms in low-resolution phase extension. *Acta Crystallogr D Biol Crystallogr* **56**, 625–33.

Wilson, K. S., Stura, E. A., Wild, D. L., Todd, R. J., Stuart, D. I., Babu, Y. S., Jenkins, J. A., Standing, T. S., Johnson, L. N., Fourme, R., Kahn, R., Gadet, A., Bartels, K. S., and Bartunik, H. D. (1983): Macromolecular crystallography with synchrotron radiation. II. Results. *J Appl Crystallogr* **16**, 28–41.

Wilson, M. A., Collins, J. L., Hod, Y., Ringe, D., and Petsko, G. A. (2003): The 1.1 Å resolution crystal structure of DJ-1, the protein mutated in autosomal recessive early onset Parkinson's disease. *Proc Nat Acad Sci USA* **100**, 9256–9261.

Wilson, W. W. (2003): Light scattering as a diagnostic for protein crystal growth – a practical approach. *J Struct Biol* **142**, 56–65.

Woolfson, M. M. (1997): *An Introduction to X-ray Crystallography, 2nd edn*, p. 161. Cambridge University Press, Cambridge, UK.

Wright, J. P., Besnard, C., Margiolaki, I., Basso, S., Camus, F., Fitch, A. N., Fox, G. C., Pattison, P., and Schiltz, M. (2008): Molecular envelopes derived from protein powder diffraction data. *J Appl Crystallogr* **41**, 329–339.

Wyckoff, R. W. G. (1948): The electron microscopy of macromolecular crystals. *Acta Crystallogr* **1**, 292–294.

Xiao, B., Spencer, J., Clements, A., Ali-Khan, N., Mittnacht, S., Broceno, C., Burghammer, M., Perrakis, A., Marmorstein, R., and Gamblin, S. J. (2003): Crystal structure of the retinoblastoma tumor suppressor protein bound to E2F and the molecular basis of its regulation. *Proc Nat Acad Sci USA* **100**, 2363–2368.

Yamada, Y., Phonda, N., Matsugaki, N., Igarashi, N., Hiraki, M., and Wakatsuki, S. (2008): Implementation of remote monitoring and diffraction evaluation systems at the Photon Factory macromolecular crystallography beamlines. *J Synch Radiat* **15**, 296–299.

Yamashita, A., Singh, S. K., Kawate, T., Jin, Y., and Gouaux, E. (2005): Crystal structure of a bacterial homologue of Na^+/Cl^--dependent neurotransmitter transporters. *Nature* **437**, 215–223.

Yang, F., Dauter, Z., and Wlodawer, A. (2000): Effects of crystal twinning on the ability to solve a macromolecular structure using multiwavelength anomalous diffraction. *Acta Crystallogr D Biol Crystallogr* **56**, 959–964.

Yano, J., Kern, J., Irrgang, K. D., Latimer, M. J., Bergmann, U., Glatzel, P., Pushkar, Y., Biesiadka, J., Loll, B., Sauer, K., Messinger, J., Zouni, A., and Yachandra, V. K. (2005): X-ray damage to the Mn4Ca complex in single crystals of photosystem II: A case study for metalloprotein crystallography. *Proc Nat Acad Sci USA* **102**, 12047–12052.

Yau, S. T., and Vekilov, P. G. (2000): Quasi-planar nucleus structure in apoferritin crystallization. *Nature* **406**, 494–7.

Ye, J. D., Tereshko, V., Frederiksen, J. K., Koide, A., Fellouse, F. A., Sidhu, S. S., Koide, S., Kossiakoff, A. A., and Piccirilli, J. A. (2008): Synthetic antibodies for specific recognition and crystallization of structured RNA. *Proc Natl Acad Sci U S A* **105**, 82–7.

Yeates, T. O. (1997): Detecting and overcoming crystal twinning. *Methods Enzymol Pt A* **276**, 344–358.

Yeates, T. O., and Fam, B. C. (1999): Protein crystals and their evil twins. *Structure with Folding & Design* **7**, R25–R29.

Yee, A. A., Savchenko, A., Ignachenko, A., Lukin, J., Xu, X. H., Skarina, T., Evdokimova, E., Liu, C. S., Semesi, A., Guido, V., Edwards, A. M., and Arrowsmith, C. H. (2005): NMR and X-ray crystallography, complementary tools in structural proteomics of small proteins. *J Am Chem Soc* **127**, 16512–16517.

Yeh, J. I., and Beale, S. I. (2007): Calorimetric approaches to characterizing effects of additives on protein crystallization. *Cryst Growth & Design* **7**, 2134–2139.

Yeh, J. I., and Hol, W. G. J. (1998): A flash-annealing technique to improve diffraction limits and lower mosaicity in crystals of glycerol kinase. *Acta Crystallogr D* **54**, 479–480.

Zheng, B., Roach, L. S., and Ismagilov, R. F. (2003): Screening of protein crystallization conditions on a microfluidic chip using nanoliter-size droplets. *J Am Chem Soc* **125**, 11170–11171.

Zheng, B., Tice, J. D., Roach, L. S., and Ismagilov, R. F. (2004): A droplet-based, composite PDMS/glass capillary microfluidic system for evaluating protein crystallization conditions by microbatch and vapor-diffusion methods with on-chip X-ray diffraction. *Angew Chem Int Ed Engl* **43**, 2508–11.

Zhou, Y., Morais-Cabral, J. H., Kaufman, A., and MacKinnon, R. (2001): Chemistry of ion coordination and hydration revealed by a K^+ channel-Fab complex at 2.0 Å resolution. *Nature* **414**, 43–8.

Zhu, X., Xu, X., and Wilson, I. A. (2008): Structure determination of the 1918 H1N1 neuraminidase from a crystal with lattice-translocation defects. *Acta Crystallogr D Biol Crystallogr* **D64**, 843–50.

Zuk, W. M., and Ward, K. B. (1991): Methods of Analysis of Protein Crystal Images. *J Cryst Growth* **110**, 148–155.

Zulauf, M., and Darcy, A. (1992): Light-Scattering of Proteins as a Criterion for Crystallization. *J Cryst Growth* **122**, 102–106.

Zwart, P. H., Grosse-Kunstleve, R. W., and Adams, P. D. (2005): Xtriage and Fest: automatic assessment of X-ray data and substructure structure factor estimation *CCP4 Newsletter* **42**; STFC, Daresbury, UK.

Index

The manufacturer's authorised representative in the EU for product safety is Oxford University Press España S.A. of el Parque Empresarial San Fernando de Henares, Avenida de Castilla, 2 – 28830 Madrid (www.oup.es/en or product.safety@oup.com). OUP España S.A. also acts as importer into Spain of products made by the manufacturer.

www.ingramcontent.com/pod-product-compliance
Lightning Source LLC
LaVergne TN
LVHW020055220926
848484LV00003BA/743
* 9 7 8 0 1 9 9 2 1 3 2 5 2 *